The Great SOUTHWEST

By Charles McCarry
Photographed by George F. Mobley

Prepared by the Special Publications Division
National Geographic Society, Washington, D. C.

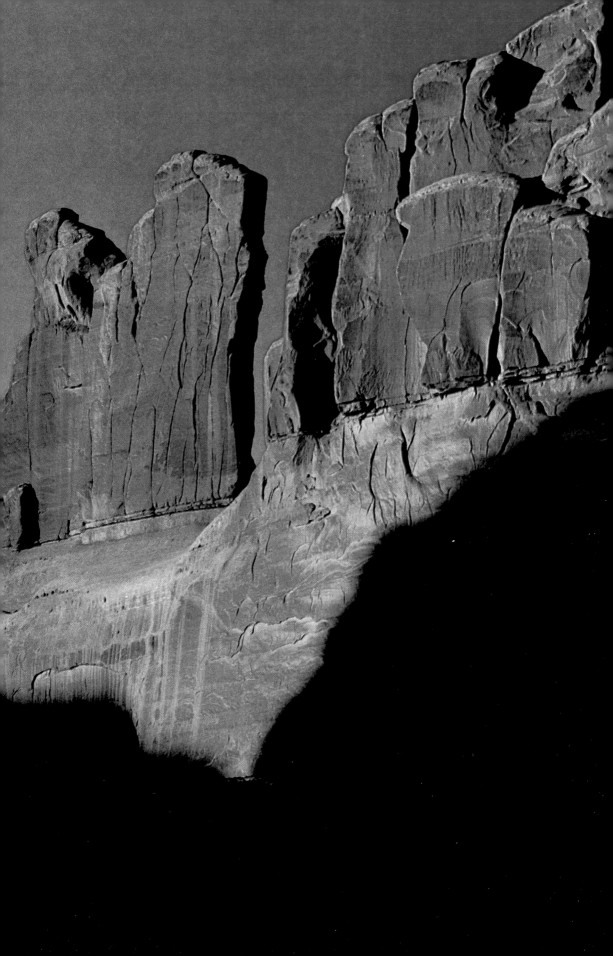

THE GREAT SOUTHWEST

By CHARLES MCCARRY
Photographed by GEORGE F. MOBLEY

Published by
The National Geographic Society
GILBERT M. GROSVENOR, *President*
MELVIN M. PAYNE, *Chairman of the Board*
OWEN R. ANDERSON, *Executive Vice President*
ROBERT L. BREEDEN, *Vice President, Publications
and Educational Media*

Prepared by
The Special Publications Division
DONALD J. CRUMP, *Editor*
PHILIP B. SILCOTT, *Associate Editor*
WILLIAM L. ALLEN, WILLIAM R. GRAY,
Senior Editors

Staff for this Book
MERRILL WINDSOR, *Managing Editor*
WILLIAM L. ALLEN, *Picture Editor*
JODY BOLT, *Art Director*
ALICE K. JABLONSKY, JENNIFER C. URQUHART,
Researchers
MONIQUE F. EINHORN, *Assistant Researcher*
Illustrations and Design
SUEZ B. KEHL, *Assistant Art Director*
CYNTHIA BREEDEN, *Assistant Designer;*
HOLLY BOWEN, *Design Assistant*
JOHN D. GARST, JR., MARGARET DEANE GRAY,
GARY M. JOHNSON, MARK H. SEIDLER,
Map Research, Design, and Production
LESLIE B. ALLEN, JANE H. BUXTON,
LOUIS DE LA HABA, JANE R. MCCAULEY,
LISA OLSON, JOSEPH REAP,
JENNIFER C. URQUHART, SUZANNE VENINO,
Picture Legend Writers
Engraving, Printing, and Product Manufacture
ROBERT W. MESSER, *Manager*
GEORGE V. WHITE, *Production Manager*
MARK R. DUNLEVY, *Production Project Manager*
JUNE L. GRAHAM, RICHARD A. MCCLURE,
RAJA D. MURSHED, CHRISTINE A. ROBERTS,
DAVID V. SHOWERS, GREGORY STORER,
Assistant Production Managers
SUSAN M. OEHLER, *Production Staff Assistant*
DEBRA A. ANTONINI, PAMELA A. BLACK,
BARBARA BRICKS, JANE H. BUXTON,
MARY ELIZABETH DAVIS, ROSAMUND GARNER,
NANCY J. HARVEY, SUZANNE J. JACOBSON,
ARTEMIS S. LAMPATHAKIS, VIRGINIA A. MCCOY,
CLEO PETROFF, ELLEN QUINN,
MARCIA ROBINSON, MARIA A. SEDILLO,
KATHERYN M. SLOCUM, PHYLLIS C. WATT,
Staff Assistants
GEORGE I. BURNESTON, III, *Index*

*Surfing on the seats of their pants,
youngsters skim down the soft dunes at
White Sands National Monument in New
Mexico, largest gypsum desert in the world.*

PAGES 6-7: *December snows mantle the Great
Sand Dunes at the base of the Sangre de
Cristo Mountains in southern Colorado.* PAGES
2-3: *Like skyscrapers flanking a city street,
sandstone formations line Park Avenue in
Arches National Park, Utah.* PAGE 1:
*Magenta blossoms of a beavertail cactus, a
member of the prickly pear family, add a spark
of color to the southern Utah landscape of Zion
National Park.* HARDCOVER: *Prickly pears
thrive throughout the Southwest.*

Foreword

I HAVE HAD A HAND in 94 National Geographic books, but this one, more than all the others, holds a special charm for me. It chronicles a sunburnt corner of our country whose geological wonders are rivaled only by the stalwart brand of people who live there, sustained by a pioneering spirit that won't go away.

For me, *The Great Southwest* is a book of memories. Both of my parents were born in Indian Territory long before Theodore Roosevelt's signature put Oklahoma's star on our flag. My mother, a quarter-blood Cheyenne, was born in a tepee, and she could remember *her* mother's frightening account of soldiers' bullets whining past during the fighting at Sand Hill in 1875, one of the last Indian skirmishes in the Territory.

This book also reminds us that the Southwest has, at times, been cruel to red man and white man alike. I grew up during the Dust Bowl days, and can recall the strange reddish brown clouds that rolled in to darken the sun. Chickens would go to roost at noon. And I can remember the exodus—battered old cars and trucks laboring past our house on U. S. Highway 66, heading west for the promised land.

Today a new and different Southwest greets my children when I take them there to visit. Drilling rigs dot the landscape. Modern farms, irrigated and fertilized, green the countryside. For several decades the people who left have been coming back, joined by thousands of newcomers. Ironically, the Southwest now finds itself the promised land.

Charles McCarry's lyrical text about this distinctive region rings true. For me it paints vivid pictures of the people, the places, the past. As I read I can hear the summer winds whisper in the cottonwoods; I can close my eyes and catch that special fragrance that only comes when rain freshens a thirsty prairie.

George Mobley's photographic genius presents a brilliant array of images from one of the richest scenic areas on the planet. His mountains soar majestically. His wild flowers duplicate nature's most extravagant palettes. His sunsets radiate a special reverence.

World traveler and author D. H. Lawrence felt such reverence when he first visited the Southwest. "It certainly has changed me for ever . . . the moment I saw the brilliant, proud morning shine high up over the desert of Santa Fe, something stood still in my soul. . . ."

There are many different ways to delineate the Southwest; the boundaries may be historical, political, sociological, climatic. The Southwest of this book coincides with the essentially arid zone that centers upon and fans out from the desert provinces of New Mexico and Arizona; it takes in most of Texas, portions of Oklahoma, Colorado, Utah, and Nevada, and the desert country of southern California.

Surprisingly, perhaps, the Southwest's shortage of rainfall has not stifled it, but rather has produced a land as diverse and beautiful as any on earth, and has bred a people as proud, as fiercely independent, yet as friendly as you'll find anywhere. The reader need not be a Southwesterner to become totally absorbed in the pages that follow.

DONALD J. CRUMP

STEPHEN TRIMBLE

Contents

Stately bison graze beneath gold-lined clouds at the
Wichita Mountains Wildlife Refuge in Oklahoma.
Established in 1905, the refuge now shelters 625 bison,
fulfilling a wish voiced by the Comanche chief
Quanah Parker: "It would make my heart glad to
see a herd once more roaming about Mount Scott."

8

"*Everyone, I think, has lived in the Southwest in his imagination....*"

AN INTRODUCTION

IT WAS THE FIRST DAY of July, but the San Juan Mountains of southern Colorado still carried last winter's snowpack on their flanks of granite and alpine tundra. Our little aircraft flew in and out of bright rainbows as the setting sun found the spectrum in rain squalls that beaded the windshield at 13,000 feet. To the north Uncompahgre Peak, driven like a flint hatchet into the American wilderness, rose high above our fragile machine.

Below us lay the Continental Divide. Not just the divide, but the place where, within reach of the eye, rise the headwaters both of the Rio Grande and of several tributaries of the Colorado River. On the eastern side, green and gently rounded, sparkling rivulets emerged from fields of snow to form little lakes, then tumbled downward, eventually to become the Rio Grande. On the steeper western slope, sun and snow sent other infant streams to join the Colorado in its 1,400-mile journey through canyon and desert to the Gulf of California.

Here is where the American Southwest begins. These alpine snows are what make the Southwest live. Totaling several hundred inches every winter, they are still melting in high summer when chocolate will dissolve in its wrapper in El Paso or El Centro. Snowmelt, dammed into great lakes, will drip from irrigation pipes and grow a crop of sweet red grapefruit for a farmer named Sam Henderson whom I came to know along the Rio Grande. Water from these same snows will push million-dollar crops toward the sunlight through the desert soil of California's Imperial Valley. Flowing through the turbines of high dams, it will make electricity to light the neon of Las Vegas and turn cameras in Hollywood, filming images of the great Southwest that will enter the minds of people all over the world, there to be stored as legend.

Everyone, I think, has lived in the Southwest in his imagination. As a boy growing up in the Berkshire Hills of Massachusetts, I would on some winter days finish my schoolwork early. When that happened, my teacher, Addie Dyer, would let me go to the unheated library that was attached to our two-room school and read. Bundled up in a woolen coat, felt boots on my feet, I would sit on the polished hemlock floor and through the clouds of my breath see mountain men and red men, cowboys and outlaws galloping along the lines of type. I may have risen every morning to milk a herd of Guernseys, and followed a slow-footed Belgian mare named Betsy through fields of timothy and clover. But in my dreams I drove Texas longhorns across the boundless plains and slept by a mesquite fire with a saddle for my pillow. In later years I was astonished to find how many boys, encountered as sober grown-ups in Berlin and Nairobi, Bombay and Tokyo, had been my unseen companions on this trail of make-believe; I even ran into a few in Dallas and Albuquerque.

The Southwest, then, is a state of mind as much as it is a geographical region. It is a generous and honest state of mind. The people of the Southwest greet you with a smile. If you ask a question, they will give you a plain answer straight from a mind that is made up. They will feed a stranger before he mentions that he is hungry, give him a drink of water because they know he must be thirsty after traveling through a dry land. And they'll talk, with humor and love, of the things that matter to them: their families, their work, their faith, and their freedom.

Back east, we talk about "the land." Out here, that is too small a term. Southwesterners speak of "the country." They would ask me, almost always, "How do you like this country?" And always I would reply, from a mind that was made up 40 years ago in that chilly Massachusetts library: "I *love* this country." For when I came here at last, I found the reality more vivid, more stirring, vaster and more American than the idea. Paper and ink, paint and film cannot capture the colors of the country, the music of the speech, the scent of snow water in the crags, of sage on the high plateau, of herds and flocks upwind in seas of prairie grass.

For generations this Eden was called a dead land, the "badlands,"

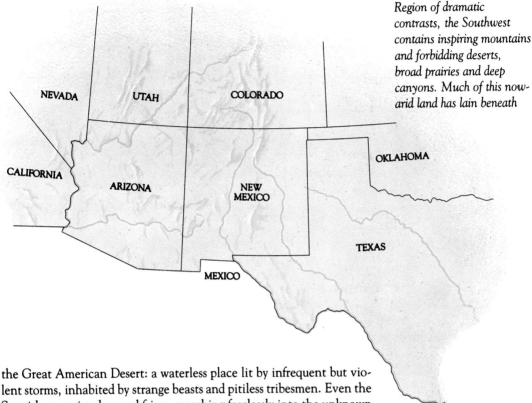

Region of dramatic contrasts, the Southwest contains inspiring mountains and forbidding deserts, broad prairies and deep canyons. Much of this now-arid land has lain beneath recurring prehistoric seas. For thousands of years, Indians lived quietly in balance with nature. Today's Southwesterners shape nature to their will. Digging mines, damming rivers, irrigating fields, building cities, they have fashioned an optimistic society that attracts thousands of new adherents every week.

the Great American Desert: a waterless place lit by infrequent but violent storms, inhabited by strange beasts and pitiless tribesmen. Even the Spanish conquistadors and friars, marching fearlessly into the unknown behind the Cross, never grasped the whole of this country. They harvested few souls among the wandering tribes who, most of them, called themselves simply "The People" and had no idea that there existed another people who would persist in regarding all the tribes as a single race called "Indians." The Peoples, who heard prayers issuing from the earth, the creatures, the plants, and the wind, had been here for thousands of years before the first Spaniards came.

When the Spaniards did come, they were lost and naked, the shipwrecked survivors of an expedition of 400 that had set out in 1528 to explore a new royal province stretching from Florida to the Rio Grande. After eight years among the tribes, who had in turn succored and enslaved them, four of the 400 found their way to a Spanish settlement in what is now northern Mexico. Three of the survivors were Spanish aristocrats; the fourth was a black Moorish slave named Esteban. Their leader, Álvar Núñez Cabeza de Vaca, repeated dramatic tales told to them by the Indians, tales of cities of fabulous wealth lying to the north.

The Seven Cities of Cibola, with turquoises (perhaps emeralds!) studding the doorframes of noble houses, beguiled the imagination of the Viceroy of New Spain. He sent an expedition led by a Franciscan, Fray Marcos de Niza, in search of the cities of treasure. Esteban guided the party, for none of the aristocrats wished to return to the interior. In

May 1539 the Moor reached the first of the Seven Cities. There, when the inhabitants fell upon him, he found not riches but death.

Returning home, Fray Marcos reported to the viceroy that he had viewed the hostile city from a distance and that it was larger than the City of Mexico. In 1540 the famous expedition led by Francisco Vásquez de Coronado found the fabled Cities of Cibola. They were six, not

Sandhill cranes take wing over Bosque del Apache National Wildlife Refuge in New Mexico. At the eastern edge of the Rio Grande Valley, a full moon shines on winter's leafless branches on the slopes of Sierra Blanca Peak near White Sands National Monument.

seven. The one in which Esteban had died was called Hawikuh. There were no emeralds in Hawikuh; like all its sister pueblos, it was a place where the greatest luxury was a strip of shade thrown on the dust by a low house made of rocks.

Seeking treasure, the Spaniards in fact brought treasure to the Southwest. In time Spanish horses transformed young tribesmen into mounted hunters and warriors. Spanish settlements were ravaged; women and children were captured and bred into the tribes. The Comanche boasted that they let the Europeans remain only because they raised such good horses for the braves to steal.

Still, the search for wealth continued to draw outsiders. Mountain men, French and American, arrived in the early 19th century, seeking the precious beaver in the mountain streams. They discovered high passes, explored rivers, and more than any other whites, merged with the land as the Peoples had done. But just as the supply of pelts for beaver-skin hats was running out, the silk hat became fashionable in London and Paris and New York, and this vivid band of adventurers vanished after its brief day.

In 1821, the year in which Mexico seized its independence from Spain, an American named Stephen F. Austin arrived in Texas at the head of a group of colonists. Austin sought the most lasting treasure of all: land. He found it with a vengeance. His Texans rebelled against Mexico, established a republic, then joined the Union, and the Mexican War ensued. By the Treaty of Guadalupe Hidalgo, signed in 1848, Mexico ceded all of its territory north of the Rio Grande—more than half of old Mexico and at least a quarter of what is now the United States—to the victorious Americans. What had been the Mexican North became the American Southwest. But Spanish-speaking folk on both sides of the Rio Grande still call that stream *Río Bravo*. And, as a third-generation native-born American told me in an elegant restaurant in Laredo, in Spanish: "This is occupied Mexico—love it or leave it."

The promise of treasure brought waves of blue-eyed people galloping into this new American outland. In Texas and Oklahoma they found free range and feral cattle and, eventually, oil. Farther west, silver

Light snow and ground fog create an eerie atmosphere at Monument Valley on the Utah-Arizona border. The

famous Mitten Buttes and other remnants of a once-broad plateau shape the valley's distinctive skyline.

and gold and copper—and, eventually, water—opened and populated New Mexico and Arizona.

All along, of course, the real treasure—usually ignored—was the heritage of the Peoples themselves. At the beginning of my journey, I knelt in a cave at the base of a cliff near the Rio Grande with a Texan named Jack Skiles and his son, Russel. In the space of seconds, my fingertip touched the tooth of a modern bison, the eye socket of its larger ancestor *Bison antiquus,* and the bones of a mammoth. In two different epochs over a span of 400 generations, men had driven bison herds over this cliff, butchered and eaten them. They killed, but they also created: In Texas and Utah and Arizona, I saw their paintings and the prints of their childlike palms aglow on the walls of other caves. Often, I felt that my hand had touched theirs across the centuries of silence and mystery that separated us. Never was this feeling stronger than on a summer's night when I had nearly completed my journey. I was camped on the South Rim of the Grand Canyon. The stars were out, and a shining object sailed through Cygnus: a satellite made by men. The handprint on the wall of the cave and the satellite moving among the stars were made, each of them, by human beings reaching as high as they could.

Much has been said and written about what man has done to the country, and how slowly the wounds he has inflicted upon it will heal. It is true that the walls of a cliff dwelling will last for a thousand years in the desert, and that the wheel tracks of a pioneer wagon train may remain for more than a century. Whether one sees these marks of man's passage as scars or reads them as glyphs containing the meaning of his existence is a matter of vision. It is well to remember that man is a part of nature and that his works, whatever they may be, are no more "unnatural" than the flight of an eagle or the passage of a tornado.

THE SOUTHWEST is a place of optimism, of confidence, of idealism. Out here they still say, with a brisk nod of the head: "If a man *wants* to do something, he can go ahead and do it." That idea is one that all Americans once had. How interesting it is that, having languished in more fertile places, it should grow again so lustily in the desert.

Perhaps that determination has something to do with the look of the country, so stark and open, and with the violent moods of its weather, its rivers, its very soil, which can grow a bumper crop in one decade and blow away into a neighboring state the next. "It'd turn noon into night," an Oklahoman told me, remembering the Dust Bowl of the 1930s. "You'd be having dinner and all of a sudden the sun would go out and somebody's farm, blown down from Colorado, would whistle through the keyhole and bury the kitchen table." This is not a place for the timid. The history of its people, red and white, has been one long struggle to make peace with the environment.

The Papago never disturbed the land by so much as the digging of a well. For generations they searched out the roots of edible desert plants, grew vegetables with the water they caught when the rains fell, and harvested the fruit of the giant saguaro cactus to make jams and ceremonial wine. Their Pima cousins, on the other hand, were skilled farmers. Living only a few miles from the Papago along the banks of rivers, they constructed irrigation canals that followed part of a complex network first designed by the prehistoric Hohokam. In turn nineteenth-century engineers, unable to improve on the ancient sites, built *their* canals on the courses laid out by the aboriginal technologists.

Today many believe that technology—which of course includes the stone ax as well as the ion beam accelerator—has, as it were, prolonged the war with the environment, and inflicted needless wounds upon it. Charles Bowden, a young scholar who is engrossed by such questions, doubts the long-term benefits of "the technological fix." It might be better, Charles suggested to me in Tucson, for man to stop pushing aside difficulties and solving problems, and to live within his ecological means. It might be better for the spirit, too. The Papago, so long as they "listened to the earth," were in harmony with nature. When they began to give up their traditional ways, life changed drastically for them as a people.

In California, a young government scientist spoke to me of the harm that might be done to invisible life—seeds, minuscule organisms—by the tread of a motorcycle tire. Do you, I asked, have data to support that statement? No, he replied; but if we do not *know*, we haven't the right to take the chance. In Texas, I helped an equally young rancher dig a deeper hole in what seemed to be a dry creek bed; soon a little moisture seeped out of the baked soil, to be sucked uphill through plastic pipe by a windmill to water young livestock. "The water's there, it's just that the creek's running underground," said the rancher. Two approaches to looking at the invisible, each in its way wonderful.

"As kids, when we'd see rain in the hills, we'd tear off our shirts and run up the wash to meet the flood," a friend told me in Arizona. That's how great a treasure water can be. The Southwest's settlers dammed the rivers to stop the floods, save the snowmelt and meager rainfall, and lift up towns where flags unfurled over school and courthouse and bells rang in church steeples. Work and water and faith filled a corner of what seemed to most white men a great emptiness. A Texan flew me over Dallas. Pointing down at the glittering city, he cried into the slipstream, "Most of us had granddaddies who came here with *nothing!*"

All the people are close to their history. If three generations ago the whites had nothing, three generations before that the red men had the freedom of a wilderness, and the Spaniards still nursed the illusion of having created a new Spain. But when, in 1540, Captain García López de Cárdenas of the column of Coronado became the first European to see the Grand Canyon, its oldest rocks had already been there for nearly two billion years, and the Colorado River had taken perhaps nine million years to reveal them by carving this stupendous cleft in the earth.

On a blistering noon 440 years later, a mere tick of the clock, I sat on Gray Mountain in northern Arizona with my boots dangling over a cliff. Three thousand feet below, the Little Colorado River raced through its serpentine gorge on the way to its junction with the Colorado in the Grand Canyon, off to the northwest.

I could see for 60 miles, to the Vermilion Cliffs, but I saw no other living creature except for a swallow or two, dozing on the breathy air. The breeze ruffled my notebook, breaking the absolute silence. Within the rock on which the notebook rested were fossils, countless shells and corals, for this was the shore of a great sea, the Kaibab, that had existed here 230 million years before.

Far away I glimpsed a plume of dust and the glint of sun on glass. It was a pickup truck, moving across the bed of an even older sea. I knew that its wheel track might last for a hundred years. And I remembered that the Toroweap Sea, which evaporated some 235 million years ago, still whispers from its rocks of the glory of the life that teemed within it.

Enigmatic figures challenge the imagination just outside Arches National Park in Utah. The images, nearly three feet tall and painted at least 1,400 years ago, survived till 1980, when vandals scrubbed away most of the paint.

FOLLOWING PAGES:
Sunset highlights Engineer Pass in Colorado's San Juan Mountains. The scene of gold and silver mining ever since Spanish colonial times, the area now bases its economy primarily on ranching and tourism.

Nearing the end of a 1,885-mile journey to the Gulf
of Mexico, the Rio Grande meanders through its
fertile floodplain. Irrigation laterals tap the river water
for bountiful crops of citrus, vegetables, and cotton.
In the semitropical lower Rio Grande Valley,
farmers count on a 365-day growing season.

"We had slept in Texas, but Mexico was only a birdsong away...."

THE LONG VALLEY

ALL THROUGH THE NIGHT, as the embers of our campfire died, the Rio Grande had whispered to us in our sleep. Its sibilant voice roused me at six. The moon, three days past the full, was as white as frost in the western sky. Camped in the shadow of cliffs 1,500 feet high, we had slept in Texas, but Mexico was only a birdsong away: As the rising sun ran a shy finger across the lip of the eastern wall of Santa Elena Canyon, I heard cliff swallows muttering as they woke in their mud nests south of the border.

Though it was full morning, the sun had not yet climbed to the top of the bluffs, and the dew was still frozen on our sleeping bags. My wife, Nancy, drinking strong coffee from the best of cups, a tin can scoured in the river sand, hugged herself for warmth and stayed close to the fire. Then the sun sent its rays through portals in the rock, so that a high butte to the northwest leaped from the darkness into the light. We breathed in the new day and heard wings beating. "Ah!" cried Nancy. "The *oneness* of it!"

The first European explorers did not quickly understand the unity of the Rio Grande and its long valley. Sixteenth-century Spaniards "discovered" the river three times and gave it three different names, never realizing that they were claiming the same body of water for the throne of Spain. Small wonder, for the last discovery was made some thousand miles upstream from the first.

Yet the Rio Grande *is* one—a great stream that rises at an altitude of nearly 13,000 feet in the San Juan Mountains of Colorado and runs 1,885 miles to its mouth on the Gulf of Mexico. In Colorado and northern New Mexico and the Big Bend country of Texas, it has cut gorges so deep that one can stand at the top and see golden eagles hunting hundreds of feet below. When it crosses the desert, it grows so shallow that a man lying on his stomach must drink carefully or his tongue will touch the bottom.

At Brownsville, where a tropical Rio Grande approaches the Gulf of Mexico, I fell into conversation with a passerby one January day, and my rusty Spanish mingled with the cry of the thousands of ravens that winter here, roosting in the palms that fringe the riverbanks. In the weeks that followed I would hear much Spanish, and meet many people, some of them American-born, to whom it was the mother tongue.

The Brownsville vicinity was the site of the first battle of the Mexican War—a conflict that still touches the daily lives of border families. "My ancestors went to sleep one night in Mexico in the year 1848, and woke up the next morning in America," explained Luciano Guajardo, director of the Laredo Public Library. "I am an American, but I can call across the river to hundreds of cousins. And I call in Spanish."

Upriver, in the thriving community of McAllen, I went out into a citrus grove at dawn to learn one of many lessons about the partnership and mutual respect that exist between the two peoples of the valley. My host was Sam Henderson, a grower of grapefruit and oranges on land irrigated by the waters of the Rio Grande. Sam has known the special satisfaction of having a new variety of grapefruit named for him. But he was quick to share credit for its discovery. "Manuel Ramirez, who's worked in this grove for more than 30 years, was the one who noticed the first tree about six years ago," Sam said. "He asked if I'd look at that tree— the fruit was yellow, but it blushed like a peach on one side."

Later that day, Norman Maxwell and Bob Rouse of the Texas Agricultural Experiment Station at Weslaco dropped by to take some photographs of Sam's trees and discuss the variety called Henderson Red. I asked Maxwell just how a new variety of grapefruit comes into being.

"Sometimes a tree—especially if it has been put under stress, perhaps during a freeze or a drought—will produce what we call a bud mutation, or sport," he explained. "The resulting bud sport limb bears somewhat different fruit. Many such mutations have occurred in this

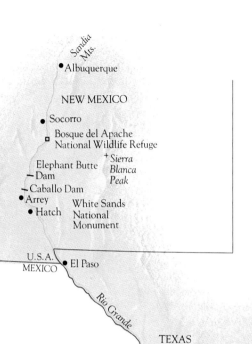

From the slopes of Colorado mountain peaks, the Rio Grande—Big River—courses southward through New Mexico. At El Paso, Texas, it becomes the international border with Mexico for 1,250 miles, threading through deep canyons and flat lowlands. Surpassed in length only by the Mississippi and Missouri among rivers of the United States, it has provided life and livelihood to this region for thousands of years. Pueblo Indians based their agricultural economy on irrigation from the Rio Grande long before Spanish colonists arrived in the 16th century.

FOLLOWING PAGES: *Ambling in long, lazy loops, the sluggish Rio Grande prepares to spill its heavy load into the gulf. Silt builds up along curves, frequently forcing the main stream into a new course and turning old channels into crescent-shaped lakes called resacas. By its devious route the river winds 55 miles from Brownsville, Texas, to its mouth—only 22 miles by air.*

valley. If the fruit has special qualities worth perpetuating, we can then reproduce the new variety by budding it onto other trees."

The first significant harvest of Texas grapefruit, with its distinctive pink flesh, was picked just after the First World War. Today grapefruit, oranges, and a few lemons, tangerines, and tangelos grow on 76,300 acres between Brownsville and Mission, a strip of irrigated land 65 miles long. A normal crop amounts to about 10 million boxes of grapefruit and 6 million boxes of oranges; each box weighs about 85 pounds.

Very little Texas citrus gets to market without passing through the sinewy hands of people who fill the quiet air of the groves with the music of Spanish. As I talked with labor contractor Ignacio Borrego in Sam Henderson's grove, both men and women, with their children tagging along and joining in the general chatter, worked steadily picking oranges. When one filled a long bag slung (Continued on page 30)   23

In Harlingen, Texas, rows on rows of trailers, each with its palm tree, house some of the thousands of "Winter Texans" who flock to the lower Rio Grande Valley to escape the northern cold. Average temperature of 75°, humidity of 60 percent, and rainfall of 25 inches mean an almost ideal climate for tourism. Like any proud homeowner, Mike Tychon (right) of Westlock, Alberta, Canada, tends his grass in Wagon City, near Mission. Tychon and his wife, Helen, look forward every year to "all the citrus we want, just for the picking," particularly grapefruit from their own tree.

Lanky fan palms outside Mission evoke images of an earlier time. When in 1519 the Spanish explorer Alonzo Alvarez de Piñeda ventured up what some historians have thought was the Rio Grande, towering palms along the banks inspired him to name the river Rio de las Palmas. Only one native grove remains. Land developers in the early 20th century planted today's palms, along with semitropical flowers and shrubs, to lure settlers from the north.

over a shoulder, he climbed a short ladder to Ignacio's truck and dumped the oranges into boxes.

"Our lives are changing," said Ignacio. "They get a little better. In 1935 I made 12½ cents an hour; now the minimum wage is $3.10 an hour, and workers can make more, depending on how much fruit they pick. In a peak period my best pickers, twin brothers, may make as much as $100 a day between them."

Ignacio's workers are Mexican Americans or Mexican nationals working legally in the States for specified periods. But the attraction of jobs also draws thousands of illegal aliens across the Rio Grande.

No one knows for certain how many such "wetbacks" cross from Mexico to the United States. "We apprehend about 780,000 illegals in a normal year," says Roger A. Stout, chief agent for the Border Patrol district administered from Laredo. "That's for the whole border, and some are repeaters—caught more than once. How many get through? More than we'd like to admit."

Once apprehended, the illegal immigrant usually agrees to voluntary repatriation—he's simply taken back to the Mexican border and released into his own country. Anyone who smuggles aliens for a fee is subject to a fine of $2,000 and a jail term of five years for each alien.

To see how the Border Patrol works, I accompanied the man who's regarded as one of the better "sign cutters" in the Laredo sector. His name is Bill Randolph, and he is supervisor of a team of Border Patrolmen who practice the old frontier skill of tracking.

"It's one of the best tools we have," said Bill. He was kneeling on a fenceline road in the wild brush country north of Laredo, touching a human footprint with a sensitive forefinger. "It's fresh," Bill said. "If it were a day old, the dew would have hardened it, or the wind moved it."

Bill and a team of three other Border Patrolmen—two on the ground, one aloft in a light aircraft—picked up the tracks of a fairly large group of men. "These fellows are pretty good," said Bill admiringly. "They're not leaving us much sign." Just then a frightened deer, a handsome whitetail, burst out of the tall brush just ahead of us and sailed over a high wire fence. "Aliens won't spook like that," said Bill. "They lie still until you touch 'em. I've walked right by a fellow curled up under a bush and never seen him."

After four hours, our patrol plane spotted a group of men lying nearly hidden in the mesquite. Bill and the two Border Patrolmen who had been tracking ahead of him closed in, running hard through the thick brush and prickly pear cactus. They came upon 11 Mexican aliens, and as Bill had predicted, none moved a muscle until a patrolman had prodded him and spoken to him. It was gentle prodding, and there was no hint of resistance. In all his 23 years of service in the Laredo area, Bill said, he'd never seen violence from men like these.

"They're just hardworking country people," said J. W. Clifford, one of Bill's sign cutters. "Good, honest folks."

When lunchtime came, the Mexicans offered to share the food they'd brought along in the plastic shopping bags that served as their luggage. Wetbacks and Border Patrolmen hunkered down together and ate chili beans and tortillas and crackers and peppers.

There is not always such easy understanding between illegals and lawmen. Some parts of the border know bitterness and violence, but in Laredo the nations are very close, and most residents are bilingual.

Don Ramiro Sanchez, descendant of Tomas Sanchez de la Barrera y

Gallardo, the Spaniard who founded Laredo, gave me a good reason. "Here," said Don Ramiro, "it is very simple—if you don't speak two languages, you don't eat."

Don Ramiro, a retired banker, had recently been created a Knight of the Order of Isabella the Catholic. This honor was awarded by the King of Spain in recognition of Don Ramiro's lifelong efforts to preserve the Spanish heritage of the people of Texas.

"Laredo is one of the largest inland ports in the United States," he told me with civic pride. "We export three to five billion dollars' worth of goods to Mexico every year through Laredo, and import two billion. We send machinery; they give us strawberries and tomatoes, fluoride and other raw materials. Business is so brisk that we have, at this moment, hundreds of boxcars waiting to cross the border. Laredo, with 100,000 people, has stores to accommodate a quarter of a million. The rest come from Mexico, to shop, to visit, to feel at home. As Mexico prospers, with its oil and other resources, Laredo will keep growing."

Nearly 85 percent of Laredo's population is of Mexican or Spanish descent. Jorge O. González, curator of history at the city's Nuevo Santander Museum, told me that this produces interesting reactions in Washington. "They ask us what we have done for minorities," said Jorge. "We reply that the Anglos—the 15 percent—are well taken care of and seem to have no complaints."

G REAT THOUGH the tradition of compromise may be in the valley, there is one force with which no compromise has ever been possible—the Rio Grande itself. Over the centuries the river has flooded repeatedly, causing terrible loss of life and property.

In 1954, when Hurricane Alice blew in from the Caribbean and dumped nearly 30 inches of rain on the Del Rio area, the river rose 56 feet in Laredo, inundating the international bridge, whose deck normally is 43 feet above the water. Upriver, the town of Eagle Pass was virtually swept away. Downstream, settlements beyond Falcon were saved by Falcon Dam, completed only months before as a joint project of Mexico and the United States.

In 1964 a huge new international dam was begun 13 miles above Del Rio. Named Amistad—"friendship" in Spanish—it was completed in 1969. Clarence Lindly, project engineer for operation and maintenance, told me how effective the six-mile-wide Amistad Dam had been in preventing disaster. "In September of 1974, we had the second largest flood of record," he said. "The inflow was 650,000 cubic feet per second. The outflow was held to 10 percent of that, preserving most of the water for constructive purposes." He patted the concrete wall, and I thought there was considerable affection in the gesture. We were inside the dam, at the very bottom. Amistad rises 254 feet above the river bed—about the height of a 25-story building—and it is capable of impounding 5,585,000 acre-feet of water.

Amistad, like Falcon Dam, is controlled by the International Boundary and Water Commission, a body that has both Mexican and American members. How is the water apportioned? "It's like a bank account," Clarence Lindly explained. "Both sides measure how much the tributaries in their countries contribute to the lake, and draw on their accounts for irrigation and other purposes."

Irrigation is an old skill along the river; the earliest Spanish visitors found the Indians diverting the river's waters onto croplands, and the conquistadors, themselves from arid country, introduced improvements. But the Amistad and Falcon Dams, in addition to preventing the devastation caused by floods and the pestilence that always follows, have greatly increased the scale of agriculture in this part of the long valley.

Amistad Reservoir has brought new prosperity to the old town of Del Rio. The reservoir abounds in game fish—at least 30 species, including bass, pike, perch, and that delight of the border palate, catfish.

Nancy thought she had encountered a fish reserved by Texans for credulous tenderfeet when Carl Cooley of Del Rio told her of catching, with bow and arrow, a fish seven feet long with jaws like an alligator and an armored reptilian body. But it turned out that Carl was an honest fisherman—he was giving an accurate description of the alligator gar, a fish that has lived in the Rio Grande since the age of the dinosaur.

With Emery (Smoky) Lehnert, a National Park Service naturalist based at Del Rio, we set out on the surface of Amistad Reservoir on a cold, windy January morning. We were bound for Seminole Canyon to look at an ancient Indian shelter called Panther Cave.

Smoky had told me that I would see a great sight, but I was unprepared for the reality. The walls of Panther Cave are covered with scores of paintings—a red panther 18 feet long, leaping deer impaled by arrows or spears, what seem to be human figures in ceremonial robes. The colors, red and black and yellow, remain vivid, preserved by the desert climate. Experts believe the paintings to be several thousand years old.

We continued up Seminole Canyon to yet another Indian cave—Fate Bell Shelter, maintained for the public by the Texas Parks and Wildlife Service. On the way down a steep trail leading to the cave, Smoky paused for a moment to extract the tough fibers of a lechugilla plant and twist them into a string. On entering Fate Bell Shelter, he stooped, picked something off the floor, and handed it to me. It was a bit of string, identical to the one he had just made—except that it had been fashioned who knows how many scores of generations ago by the brown hands whose imprints I saw, no larger than the hands of children, stained in vermilion on the walls of the cave.

I found that my fascination with those ancient cave dwellers is shared by contemporary residents of the region as I roamed the desert with a quiet 23-year-old named Russel Skiles, who had just graduated with honors from Angelo State University with a major in journalism.

Russel regards the outside world with mixed feelings. He grew up on a sheep ranch his grandfather started near Langtry, Texas. Russel's father and mother, Jack and Wilmuth Skiles, still live on the ranch and run another close by. It's a small spread by Texas standards—"only" 7,000 acres, on which some 900 ewes graze on the sparse vegetation. Russel is as much a creature of this hard country as the golden eagles that in one drought year helped to destroy 80 percent of the Skileses' crop of lambs. But there are few opportunities here. Langtry, with no major town closer than 60 miles, has a population of 39. It lives on a legend—Jack Skiles is director of a fine museum that preserves the history of Judge Roy Bean, justice of the peace, saloon keeper, and frontier scoundrel who was for a time in the 19th century "the law west of the Pecos."

Wilmuth Skiles teaches five subjects at Comstock High School, 28 miles away. There are only five boys on the basketball squad. When players foul out, the game continues; and the week I was there, they had held the score close against a bigger school despite the fact that Comstock had only two players on the floor. The opposing coach had tried to remove some of his own players, but the referee wouldn't allow it. The rules of basketball forbid such gallantries.

Jack Skiles, a tall, spare man who has a master's degree in botany and is a former superintendent of schools, welcomed me to his comfortable house on the bluffs above the Rio Grande. "You're just in time for lunch," he said. "How about some venison chili?" Jack took a big pot of this delicacy out of the freezer, popped it into a microwave oven, and in minutes I was spooning up the delicious mixture—and pondering the remarkable qualities of people who can combine venison and microwaves and come up with easy, natural hospitality.

Jack Skiles's strongest expletive is "*Chihuahua!*" To a man who has spent his life in the Chihuahuan Desert, which covers nearly 140,000 square miles in Mexico and the U. S., there is hardly a word more vibrant with respect and love for the power and wonder of nature. Jack and his collection of things he has found on his land—artifacts, relics, fossils—taught me a great deal about the desert. Just by forcing this country to let him live on it, Jack Skiles has become archaeologist, historian, botanist, biologist, teacher, rancher—a Renaissance man who dwells in a land whose colossal sculptures were carved not by Michelangelo but by the patient, irresistible chisel of the Rio Grande.

A FEW DAYS LATER, on the back of a shuffling white horse named Smuggler, I thought of Jack as I gazed down from the heights of the Chisos Mountains on some of the river's most impressive works. We were in Big Bend National Park, in the company of people who love it for its grandiose beauty and its deep silences—park naturalist Frank Deckert and his wife, Gloria. Frank and Gloria have lived in wilderness most of the years of their marriage. "We'd been here in Big Bend for a couple of years when our 2½-year-old daughter came running into the house, terrified by something she'd seen in the sky," said Gloria. "No, not an eagle—she was used to hawks and eagles. It was an airplane."

Big Bend, more than a thousand square miles in southwestern Texas where the Rio Grande makes its "big bend" through Santa Elena, Mariscal, and Boquillas Canyons, was designated a national park in 1944. We began our trip on horseback at the usual western hour of departure—sunup—and as our mounts climbed into the heights of the Chisos Mountains, we rode through a herd of six Del Carmen whitetails. The deer raised their great oval ears and gazed at us with the innocent curiosity for which they are famous. Our trail took us over rocks and cactus to the South Rim of the Chisos range. From there we gazed straight down some 2,500 feet to the desert floor. Twenty miles away and 4,000 feet below, we could see Santa Elena Canyon, where we had camped so pleasantly, and beyond it the craggy brown face of Mexico.

Frank Deckert knows almost every plant and animal in the park— the candelilla shrub, source of commercial wax; the Colima warbler, whose only U. S. nesting ground is in these mountains; the venomous brown recluse spider, the desert centipede, the Mexican jay and blackcrested titmouse that shared our lunch in the shade of an oak tree.

Moving in single file, with only the clatter of the horses' hooves on loose stones disturbing the silence, we were as alone—and as tightly bound to nature by its overwhelming power—as any band of Mescalero Apaches who ever moved over this trail.

The road along the Rio Grande ends west of Presidio, so we drove on to El Paso behind a great shield of mountains. *(Continued on page 40)*

As if with a comforting hug, Russel Skiles steadies a lamb for his father to "mark"—castrate, cut off the tail, and mark an ear as identification. The Skileses run 900 ewes on their 7,000-acre ranch near Langtry. Well suited for sheep ranching, the dry climate of southwest Texas inhibits animal parasites and supports a scrub grass adequate for sheep and goats though not for cattle.

Ragged peaks of the Chisos Mountains pierce the morning mist in Big Bend National Park. A great crook in the

Rio Grande curves around a wilderness of steep canyons, desert basin, and wooded heights.

Red chili peppers drying in the sun carpet a hillside in Arrey, New Mexico. Rosa Maria and Serjio Cardona turn and sort the chilis by hand to protect their highly prized flavor. Because of the extra care, time, and space required for sun-drying, most farmers today use modern dehydration plants. Even on large acreages, such as that of Ray Enriquez in the Mesilla Valley (above), workers still harvest by hand, starting in August and continuing well into cold weather. New Mexico has become the nation's top producer of chili peppers. More than a dozen varieties go into wine, jelly, candy, soup, food coloring, and cosmetics, as well as the traditional seasonings for Mexican-American cuisine.

Here Pancho Villa used to eat ice cream at the Elite Confectionery while visiting between revolutions. El Paso has always been important militarily, and Fort Bliss, established at the end of the Mexican War, still thrives—now as the largest air defense training center in the free world.

There has been a heavy traffic in heroes through this country. All along the Rio Grande Valley lie the ruins and legends of frontier posts. Robert E. Lee, John J. Pershing, George S. Patton, and a host of other eminent soldiers served on the border.

Traffic of other kinds across the river is a historic industry. Anything that can be smuggled has been smuggled across the Rio Grande—and certain characters who, in the old days, drove such harmless contraband as herds of longhorns and saddle horses across the lonelier fords established fortunes and reputations that still stir chuckles of admiration when men sit around in the shade and swap tales.

But there has always been a dark side to the illicit business of smuggling, and I learned in El Paso that it has never been darker than in recent years. Law-enforcement officials are losing the battle against a huge army of drug traffickers. "If things keep on this way," one federal agent told me, "marijuana—and worse—will soon be available to every schoolchild in the United States."

A joint U. S.-Mexican program of spraying narcotic plants with herbicides much reduced the production of opium and marijuana by Mexican growers. But criminals promptly found plentiful new supplies elsewhere, and now whole shiploads of marijuana are being intercepted in American waters by the U. S. Coast Guard. Vast quantities of drugs come into the U. S. by airplane—so much, in fact, that there were 180 air crashes in 1979 that officials know were related to the drug traffic.

Who is behind this enormous enterprise? "Fifteen years ago it was organized crime. Now any amateur can get into the business. If we were just dealing with the syndicates, our job would be easier."

The speaker was Jerry Posey, an official of the El Paso Intelligence Center, a federal installation that uses space-age technology to gather information on the drug trade throughout the world. EPIC is staffed by personnel from eight law-enforcement agencies; around the clock they monitor sophisticated computers and communications gear to keep watch on known violators and to identify new ones.

In 1979, federal officers seized 1,244,000 pounds of marijuana, 386 pounds of cocaine, and 208 pounds of heroin, worth in total more than half a billion dollars at street retail prices.

"Nobody knows how much gets through," Jerry Posey told me. "But we do know we're intercepting only a small part of the traffic."

Henry Washington, agent-in-charge of the El Paso District of the Drug Enforcement Administration, has spent 27 years as a Border Patrolman, Customs officer, and DEA official. He described the problem in one sentence: "If I had 500 agents in El Paso—and I actually have fewer than 50—I couldn't control the flow of drugs into this one city."

JUST NORTH OF EL PASO, the Rio Grande ceases to be the border. The traveler approaching from the south through expanses of mesquite and grama grass will sense that he is being beckoned into the interior of New Mexico by mountains. The earth rises almost imperceptibly, the air freshens, the juniper, buffalo grass, and piñon pine begin to appear. Ahead lie the blue heights of Jornada del Muerto, Sierra Blanca, the Manzanos, the Sandias, and finally the Sangre de Cristo range under its wimple of virgin snow. The elevation at the border just above El Paso is about 3,700 feet; the Sangre de Cristos rise above 13,000.

The Spaniards, like the Indians in their pueblos before them, clustered along the river. In the 18th and 19th centuries more warlike tribes—Apache who had migrated earlier from the Great Plains, Comanche newcomers from Wyoming—ranged the valley and harassed the colonists. War parties of both those nomadic groups stole women and children and horses. Once taken by the Comanche, whose great war trail swept through much of what is now New Mexico, Oklahoma, and west Texas and on into old Mexico, a captive did well to accept his fate, for his captors moved ceaselessly through a vast prison of space.

"I am a typical Comanche," says LaDonna Harris of Albuquerque, wife of former U. S. Senator and presidential candidate Fred Harris (D.-Oklahoma). "My great-grandfather was a Spaniard captured at the age of 7. One of my grandmothers was part Crow, and the other was a green-eyed, red-haired woman whose father was a Mexican."

Her blue eyes flashed with pride as she spoke of her great-grandfather: "He became a Comanche war chief, and when he was captured by the U. S. Army he was put in chains at Fort Sill, Oklahoma. The soldiers asked if he wished to be returned to his people—the Mexicans. He replied, 'I am a Comanche.' Had they struck off his chains he would have raided Mexico!"

On the Plains of San Augustin 50 miles west of Socorro, where La-Donna Harris's ancestors may well have studied the stars, I visited a place where men and women of today spend their time pondering the mysteries of the universe. With the laconic precision of science, the most sophisticated radio telescope ever constructed is called the Very Large Array. When completed in late 1980, VLA will deploy 27 radio antennas, each a dish 25 meters (about 82 feet) in diameter, on nearly 38 miles of railroad track arranged in the shape of a Y. With all its dishes operating, VLA will be as powerful as a single antenna 17 miles in diameter.

VLA has already transported human consciousness to a multiple quasar, or quasi-stellar object, 15 billion light-years away (light travels about six trillion miles in a year). My guide, a cheery young man named Jonathan Spargo, twiddled with the VLA's computerized image-processing system and brought onto the screen a picture created from radio waves. "This quasar," he explained, "is moving away from the earth at two-thirds the speed of light. The quasar is roughly one light-year in diameter, yet it emits more energy than our entire galaxy, the Milky Way, which is 100,000 light-years in diameter. That is very puzzling, because our physics can't explain so much energy from a source of that size."

He switched to another picture, this time of a galaxy 300,000 light-years across. "How many stars does that have? We don't know, but our own Milky Way, one-third the diameter of that galaxy, has two hundred billion stars," said Jonathan. "What is really out there, the truth for which VLA is searching, is very much stranger than any science fiction ever imagined by man."

In New Mexico, the Rio Grande Valley has been from the days of prehistory a stage on which technology has been displayed. The Indians of the pueblos, creating their cities and their works of art and artisanship, were a technological civilization. The Toledo blade that slashed the waters of the Rio Grande and claimed them for the throne of Spain was, in its time, among the greatest achievements of European technology. Yankee traders brought musical instruments over the Santa Fe

Trail, and Texans brought repeating six-guns up the valley from the south. The first nuclear bomb was built and detonated in New Mexico, and Robert H. Goddard moved to this state from Massachusetts to perform experiments in rocketry that laid the basis for the intercontinental ballistic missile and for interplanetary travel.

Albuquerque, a charming mixture of old adobe and sparkling new glass, is a center of modern technology and home of Sandia National Laboratories, an engineering and research institution operated by Western Electric for the U. S. Government. So important is this organization to research and development in weapons and energy that $1 of every $1,000 spent by the federal government is spent at Sandia.

"Albuquerque has a remarkably high percentage of Ph.D's because of the concentration of research here," said Dr. Lee B. Zink, a professor and administrator at the University of New Mexico who is president-elect of the Greater Albuquerque Chamber of Commerce. "This city is a brain bank for the nation."

Something in the air here, something specially optimistic and American, leads one to believe that any problem can be solved. Perhaps it is partly the sense of space. Says Tony Hillerman, an Albuquerque novelist and teacher: "The sky doesn't come down on your forehead, the way it does even in the Texas Panhandle. You hear the poetry in the landscape. Over time, it makes the spirit robust."

It has ever been a welcoming sky. Under it I made friends with an ebullient physicist named Yu Hak (Haggie) Hahn, Ph.D. Haggie was born in Korea, lost his father in the war, and came to the United States as a teenager. His mother, three brothers, and two sisters followed. The children worked their way through college and won several fellowships to graduate school; among them they have earned four master's degrees and three doctorates. "This country!" says Haggie. "It's given me the greatest gifts: education and optimism!"

Hahn, a specialist in optical components for laser instruments, founded CVI Laser Corporation in Albuquerque in 1972. He had little money but much imagination. Today he is a millionaire. In his laboratory, he showed me what had made him one—a special lens polished and coated by high technology. Looking into it, I glimpsed unearthly colors—it was no longer glass, not yet a jewel.

"We use equipment we design and build ourselves," said Haggie. "Much of what we do is in the invisible region, and no one does it as we do." He grinned. "Even the *Japanese* have to buy from us!"

Before leaving Albuquerque's world of technology, I made a pilgrimage to the Sandia laboratories. Here many of the nuclear weapons in the U. S. arsenal have been developed. I was interested in another aspect of nuclear physics—energy from fusion, the harnessing for human progress of the forces within the hydrogen bomb and the sun itself. At Sandia, a young physicist named Glenn W. Kuswa is heading a team that has already crossed this far frontier of applied science.

"Fuel for the generation of power by fusion does not enter into the cost equation," he told me. "The raw materials are deuterium and tritium, isotopes of hydrogen that are easy to find or easy to make. The technology is complicated—but that's a challenge, not a barrier."

We were standing inside the great shell of the machine—an ion beam accelerator developed at Sandia—that may lead to the first commercial fusion plant by the year 2000. Kuswa had shown me a pellet, resembling a very small pearl, that contained tritium and deuterium.

"The purpose of this device, which will cost nearly 30 million dollars," said Glenn, "is to direct an ion beam with an intensity of 100 trillion watts of electricity onto the pellet for 20 billionths of a second."

One hundred trillion watts of energy? "Yes," replied Glenn. "That's more than the electrical generating capacity of all the power stations on earth. But of course it will only last for 20 billionths of a second. We hope to do that in 1986. We've already run preliminary tests in which we produced 20 trillion watts. That lasted for a long time—almost 40 billionths of a second."

The harnessing of fusion-created energy through the use of ion beams is a new concept. Continuing attempts involving lasers, electron beams, and magnetic fields have not produced the virtually limitless power that fusion has always promised. Does Glenn Kuswa really think that this new experiment will lead to designs for commercially feasible stations in only 20 years? "I do, yes," he responded. "And when we succeed, the price of petroleum ought to drop through the floor."

NOT LONG AFTER my stop at Sandia, I visited Bosque del Apache National Wildlife Refuge, 90 miles down the Rio Grande from Albuquerque. Here another aspect of technology seeks to preserve one of nature's most splendid creations, the endangered whooping crane.

"There are now 117 whooping cranes in the world," said Ron Perry of the U. S. Fish and Wildlife Service. "That's an increase from 21 in 1941—the low point reached by this species in the wild."

In 1975, American and Canadian scientists placed whooping crane eggs in nests of the greater sandhill crane, which had itself once been an endangered species. Scientists hoped that the whooping crane chicks would be accepted by the greater sandhill cranes, and that they would adopt the much shorter migratory route of their foster parents. "It seems to be working," Perry told me in his soft Colorado accent. "We had 15 whoopers here on the refuge this year—Ida, Homer, Ernie, Dick, Jane, Ray. We know them by name, of course. We hope they'll soon be proud parents, so we can name the grandchildren."

It was nearing sunset as I wandered through the refuge, waiting for the birds to return from their feeding grounds. Clouds of blackbirds spilled from the cirrus that streamed over the Jornada del Muerto; a great blue heron lifted itself into the resisting air and, finding grace in the wind, danced out of the realm of earthbound creatures.

Soon we heard the first cries of the returning waterfowl in their thousands—Canada geese and snow geese, mallard and teal and pintails, flight after flight, black against the darkening sky, lifting their myriad voices in joyous song.

At last, in a line of greater sandhill cranes, I saw a white bird, larger than the others, pure in hue, glorious in form. A human hand had held this whooping crane in its egg and made it possible for it, and perhaps for its imperiled kind, to abide on this planet a while longer. That seemed to me as wondrous a thing as any of the works of man I had seen in the long valley—from the cave paintings of vanished Indians to the captured light of galaxies focused for my eye after a journey of trillions of miles.

The whooping crane settled into the water a few hundred feet away. And as I watched it, I was glad and proud to be what the artists, the explorers, the settlers, the warriors, the builders, and the scientists along the Rio Grande had been, each in his way and each in his age: a human being, standing in awe of nature.

FOLLOWING PAGES:
Project manager John Lancaster stands in the main reflector of one of the 27 antennas that comprise the Very Large Array radio telescope west of Socorro, New Mexico. Set on railroad tracks laid out in a Y, the movable antennas produce results equal to that of one dish 17 miles in diameter. Its site on the Plains of San Augustin offers astronomers advantages of latitude and elevation, dry climate, and a protective ring of mountains.

Abundant sunshine and open space create a prime setting for a solar thermal test facility (below) at Albuquerque's Sandia National Laboratories. Controlled by computer, 222 heliostats track the sun and reflect its energy to a receiver atop a tower. Other energy experiments at Sandia focus on use of particle beams for controlled thermonuclear fusion, an outgrowth of studies on radiation effects of nuclear weapons. Researchers have developed an accelerator (left, as viewed looking upward during construction) that will test the feasibility of fusion-created energy for future commercial use.

FOLLOWING PAGES: Lifted by propane-heated air, "The Big One" joins a mass ascent at Albuquerque's Annual International Balloon Fiesta in 1979.

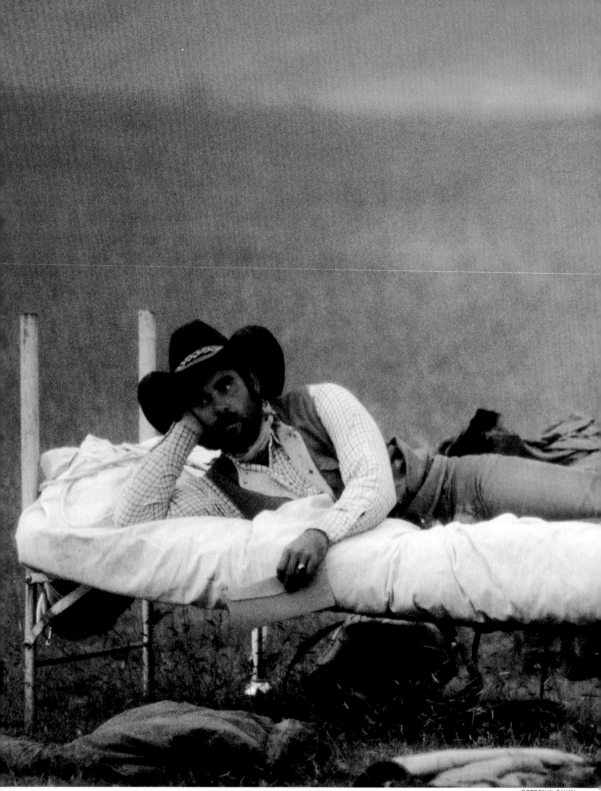

GORDON W. GAHAN

Bone-weary after a long day working cattle, a cowhand beds down on the open range of the 06 Ranch in west Texas. For more than a century the American cowboy, blending fact and legend to create his own colorful mystique, has remained an enduring figure of the great Southwest.

50

"Vast seas of hardy grass and fine dust, of bitter winds and uncertain rain...."

PLAINS AND PRAIRIES

THE COWHAND, sitting easy in the saddle of a magnificent quarter horse, wheeled at the end of a ballet between horse and whiteface calf and loped over to the corral fence where I sat watching. "I just love the sky out here," said the rider, Diane Lacy, of the 06 Ranch in the Davis Mountains of west Texas. She looked about at a twilight filled with thunderheads. "There's so much of it!"

We smiled in agreement as we admired the changing light of the dying day. Then Diane turned her mount and with a signal I could not detect—for this was a cutting horse, trained to respond instantly to movements of hand and heel and knee that are as subtle as a caress—she returned to the complicated business of making a yearling Hereford yield to the will of horse and human. It was a practice session; had it been roundup time, the calf would have been thrown, branded, earmarked, dehorned, castrated, immunized against disease—and thereby transformed from a creature that had been born wild and lived wild under the vast sky of this rolling rangeland, into a domestic animal to be fattened, slaughtered, and singed in charcoal smoke in some suburban backyard.

Melancholy thought? Well, it was a sweetly melancholy time of day. Meadowlarks raised their evensong, and the clouds above the blue rim of hills turned red with the setting sun. Then the full moon appeared on the horizon, and horses, riders, and cattle were engraved upon it like figures on a coin.

Back on the Rio Grande, as we paused at the same hour in the solitude of Eagle's Nest Canyon, Jack Skiles had told me that when I reached this country I would find the West much as the West used to be. True enough. But it has changed, and Diane Lacy—mother of two, dancer, choreographer, and painter, and a top hand on the 06 Ranch, one of the largest privately owned spreads in Texas—was the perfect proof of it. As a 21-year-old bride, she came to the ranch with her husband, Chris, whose family has owned the 06 for four generations.

"When I first arrived, there was a feeling going all the way back, I guess, to the olden times, that women just didn't belong at a roundup," Diane told me. "Then one spring day they were real shorthanded, and Chris gave me a branding iron and told me to brand. That's one of the trickiest chores of all in working cattle—getting a good, clear brand. I had a lot of people telling me about my mistakes, so I learned fast. Now it's my job—I can put brands on 250 calves a day, and get 'em all right."

Becoming a cowhand was no easy business for Diane Lacy. Breaking a colt, she broke her wrist. She had to get used to long, rugged days on three or four hours' sleep. "When we're working—and I'm talking about dawn till dark with the cattle, and then maybe fighting a grass fire by moonlight—I'm just another hand on a horse," she said. "At least I hope that's the way the men look at it."

Still, when you think about it, there's nothing very strange in what Diane Lacy has accomplished. The plains and prairies—those vast seas of hardy grass and fine dust, of bitter winds and uncertain rain—have always been a proving ground for human beings. Here animals lived in herds and red men lived in tribes, huddling against the forbidding distances. It was the cowboy—first the *vaquero*, then the *norteamericano*—who was the first solitary in this wide-open country. And the cowboy's legend insists that nobody asked his name, or where he was from, or why he was here. It was what he could *do*, by the gruff standards of men who expressed their love for this vast land and its wild herds by turning each part of their daily work into an art—roping, branding, riding, cutting, or just enduring the spiteful moods of that endless sky.

Diane simply did what a thousand slim young strangers had done before her since the days of the longhorns: She rode into camp, stayed at the edge of the firelight, and at the right time showed the grizzled old hands who knew all the tricks and all the jokes that she could do the job. There was one difference: They called her "ma'am" instead of "kid."

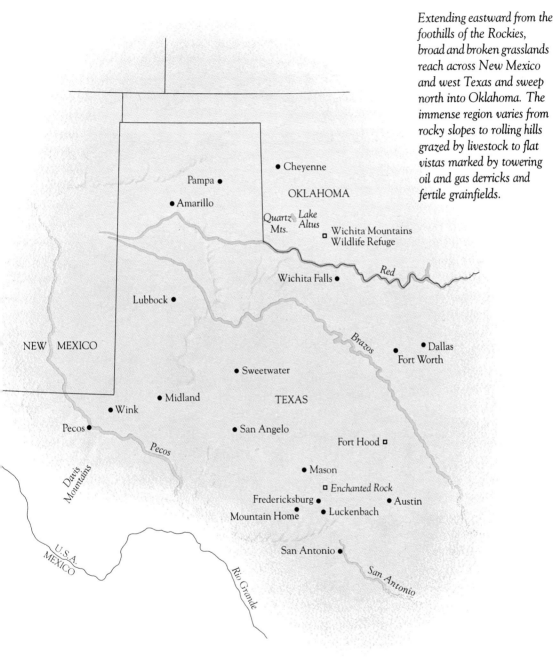

Extending eastward from the foothills of the Rockies, broad and broken grasslands reach across New Mexico and west Texas and sweep north into Oklahoma. The immense region varies from rocky slopes to rolling hills grazed by livestock to flat vistas marked by towering oil and gas derricks and fertile grainfields.

About the only thing that's close out here is history. Several hundred miles north of the 06, in the Oklahoma Panhandle, 75-year-old Truman Tucker shared a soda pop with me and got to talking about the last roundup of the Prairie Cattle Co. "The cowboys camped on our place," said Truman, "and two of 'em, Jess Corbin and his brother, were Kit Carson's nephews. That was 68 years ago." *(Continued on page 60)*

*Fall roundup on the
06 Ranch brings grueling
days and short nights.
Horses still prevail
over vehicles on the 200-
square-mile spread in the
Davis Mountains.*

At far left, ranch manager Chris Lacy, his wife, Diane, and their children, Kristin and Lance, prepare for an early morning ride. "It's taken a long time for me to be accepted as a real hand by the cowboys," says Diane. But Chris insists, "She's gotten so good we couldn't manage without her." At left, wagon boss Jack Phariss discusses the roundup with Chris; Lance, 3, has called it a day.

One verdant patch of grass marks the massive slope of Enchanted Rock, a granite dome that rises to an elevation of 1,825 feet near Fredericksburg, Texas. The huge batholith formed more than 800 million years ago when molten rock cooled below the earth's surface. Erosion later exposed and smoothed its sides and top. Comanche Indians, mystified by the sounds issuing from the rock during expansion or contraction, gave it its name.

Craftsmanship of 19th-century European settlers permeates this Queen Anne-style mansion in Mason, Texas. Built in the 1880s, the three-story house contains hand-carved 24-inch-thick stone walls, elaborate woodwork, and 48 limestone Ionic pillars. Garner Seaquist has lived here since his father, a Swedish immigrant, bought the house in 1919. At lower right, Clara and Garner Seaquist chat with their granddaughter Sandee and her cousin Gunnar in the kitchen, warmed by the wood-burning cookstove Clara still uses. A son, Michael, pauses in the silence of the closet-size retreat the family calls the "reflectory" as late-afternoon sunlight filters through its stained-glass windows.

Shaking hands, I often touched the old West. Rufus Van-Zandt, 85 years old, still takes hold of a stranger's hand as if it were the walnut butt of a Colt Peacemaker. When Rufus was a small boy in Fort Worth, then Camp Worth, a riot broke out, and the local authorities telegraphed Austin and asked for a company of Texas Rangers. "The next day," recalled Rufus, "an engine came barreling into the station, and everybody was down there to greet the Rangers. This little gray-headed fellow stepped off the engine and came trotting up the platform. He had two pistols and a sawed-off Winchester. He said to the mayor, 'I'm Capt. Bill McDonald.' The mayor said, 'Captain, where is your company?' Bill said, 'Hell, mayor, you ain't got but one riot.' Pretty soon we didn't have any riot at all. That was the old rule: one riot, one Ranger."

Capt. Bill McDonald, with his long white mustache, remained a vivid figure in Rufus Van-Zandt's mind. In 1922, after Army service on the Mexican border and in France, Rufus joined the Texas Rangers. "It was just at the time of the national railroad strike," he told me. "We had the railroads to take care of along with the bootlegging and the cattle rustling and everything else. My time in the Ranger service was a little bit on the wild side. You could get a gunfight anytime you wanted it. Kill a man and there is always somebody that wants to see if he can't beat you to the draw."

Being fast with a gun was the only way to stay alive in Texas even in the 1920s? "In that line of work, Lord, yes," said Rufus. "If you were in Texarkana and killed a man, why, some other fellow would come on up from Dallas or El Paso to see if he couldn't run into you and challenge you. It was sport: 'I'm the man that killed McDonald or Van-Zandt.' Live by the gun and die by the gun."

When Rufus Van-Zandt became a Texas Ranger, the west Texas oil fields were booming, and in Rufus's words "they were pretty rank." That, as I discovered when I called on a wildcatter named Ford Chapman, was a matter of opinion. Ford lives in Midland, Texas, a sort of Brasilia of the plains, whose towers were raised toward the sun by one thing—oil. The Texas oil boom began in Corsicana about 1895, and the fabulous Spindletop gusher was brought in six years later.

As a small boy, just after the Great War, Ford traveled with his father through rural Texas selling drill sites. A promoter like Ford's dad would buy a section of land (one square mile, 640 acres) for three or four dollars an acre, subdivide it into one-acre drill sites, and sell the sites for $100 apiece. "It was a punchboard operation," said Ford. "The odds were a million to one against finding oil, but if you did. . . !"

By the time he became a teenager, Ford was working behind a soda fountain in a drugstore and began to notice the oil men who hung around the place: "They all had good cars and they were optimists. I decided that was the business I was going to get into."

He moved into his first tent in an oil field on June 11, 1930. In the middle of November a gusher shot out of the hole in the punchboard, and Ford Chapman, working a 12-hour day for board and a promise, was hooked. An older oil man who hired him as a roustabout lifted the boy's hat and tapped him on the skull. "Use your head," he advised. "I sure will," Ford replied. And he sure has. After the Second World War, Chapman bought a wooden rig and went into business as a wildcatter. With a few dry holes and a little success behind him, he bought two more rigs. In 1948, west of Pecos, he traded a man 80 acres for an extra rig and brought in a well—a big one.

"*Whew!*" said Chapman, seeing the gusher in the mirror of his memory. "Look at that baby shooting over the top of the rig—a solid stream of oil eight inches thick. I used to run through that black shower just to smell it and feel it hit me on the skin. You ever do it once, you gotta do it again!" Ford Chapman did it again and again.

"Today we have the advantage of modern science," he said. "But years ago, me and the others, we played hunches. You could get to where you could *feel* oil. You'd hit a dry hole, move your rig one mile, and hit millions of barrels."

A look of manly joy comes into Ford Chapman's eyes when he speaks of those times. "They were the happiest of my life, and I wouldn't have lived 'em any other way. A town like Wink was open 24 hours a day—lights burning, car horns blowing, music making a racket. Gamblers. Bootleggers. Red lights—they used to say a boom had started when the scarlet women came, and ended when they drifted off.

"You didn't know many folks by their full names; there was no government much in those days, and nobody kept up with you. I knew men for years, worked with 'em, made deals with 'em by word without a scratch of a pen, and never knew their last names."

Wink was located in one of the 89 counties in which Rufus Van-Zandt defended the law. Once a bootlegger named Blackie Robinson set Rufus up for an ambush, but Rufus ambushed the ambushers. Then he took Blackie to an empty cattle tank near the New Mexico line and tied him in with fence wire. "I figured that was a good place for him," said Rufus, "just like he figured dead was a good place for me. The sheriff said, 'Captain, you're a mean one.' I told him, 'No, I'm just playful.'"

Playfulness is one of the bright colors in this boundless, often monotonous landscape. Van-Zandt once had a mountain lion as a pet: "I caught him when he was still a bleary-eyed kitten and his coat was spotted like a fawn's. I used to fly, and I'd carry him in my airplane to New York or Chicago, strapped in with a special collar. He'd sit up in the front cockpit with his mouth open and his whiskers blowing back, looking all around. He was the nicest cat I ever saw."

Search for topaz, the state gem of Texas, leads collectors to Mason County the year round. Stones such as the pale blue specimen above—nearly three inches long—form in pockets in granite. Slow wearing of the rock by streams and rainwater releases the embedded crystals. Behind the stone, a hedgehog cactus displays its scarlet bloom.

SOME SAGE has observed that plainsmen, and Texans in particular, like contests: fiddling contests, chili-eating contests, mudwallowing contests, and rodeos—once described to me as "falling down contests." Sam Lewis of San Angelo is one of the originators of the World Championship Armadillo Race (the record time is 3.5 seconds over a 40-foot course), the World Championship Rocky Mountain Oyster Cookoff, and the World Championship Chili Cookoff.

The boys like to fun Ol' Sam just as they like to fun everybody, and at the chili cookoff at Terlingua they put a five-foot rattlesnake in his bedroll. "It was one of them rattlesnakes where they sew the jaws together so's all it can do is stick out its tongue," said Ol' Sam. "But I didn't *know* that right away."

At a brisket cookoff in Menard, Texas, I ran into a fellow who was able to tell me something I'd always wanted to know: how high a rattlesnake can bite. "Depends on the snake," said Glenn Wortham of the J2 Ranch at Sweetwater, where an annual rattlesnake roundup is held. "A 70-inch western diamondback will hit an ordinary-size fellow about mid-thigh. But that's a big snake for out here. Biggest we ever caught on the J2 was 66 inches, and 83 inches was the biggest I ever saw."

I was too late for the cow-chip throwing contest (claimed record:

273 feet) but I did run into John Lee Sawyer of San Angelo, the world champion tobacco spitter. "I won the title last year in Sonora, Texas, with 36 feet 1 inch," bellowed John Lee over the screech of a children's fiddling contest. He has won 17 trophies and has never been defeated. "The secret is T. J. Swan Tobacco, this here split between my teeth, and compression," said he, laying a ringed hand on a noble paunch.

On a spring evening a male axis deer bounds through a woodland on the TBT Ranch near Mountain Home, Texas. The 2,000-acre property provides a home and breeding ground for 14 species of rare and endangered animals. Moderate climate and large tracts of uncultivated land, much of it unsuitable for profitable livestock operations, have spurred the development of game ranches in this central Texas region.

Nearby, four serious men were tossing 2 ½-inch metal washers in high arcs. The object was to drop them into a cup buried in the dirt 21 feet away. "What do you call that game?" I asked. "Warshers," said one of the men. "Goldarn!" hollered a player as he missed a ringer, losing the match. He lifted his boot off the $100 bill he'd been keeping from blowing away, and his opponent picked it up.

THESE FOLKS have earned a little fun. There were years on these bleak lands when only the blood on the ground kept the dirt from blowing away. For hundreds of years, until Samuel Colt invented the six-shooter, the white invader was poorly armed in comparison to the Plains Indian. From the back of a running pony, the Comanche warrior could drive an arrow entirely through the body of a buffalo. A skilled bowman could keep a steady stream of arrows in the air. So great was the military superiority of the Indians that Apache raiders cut off all access to Laredo at one point, and the Comanche with their long lances and short bows made sport of the Texas Rangers until the battle of the Pedernales in 1844, when they encountered revolver fire for the first time.

It was the Spanish conquistador and his successors who made the Plains Indian into the great cavalry fighter he became. When Coronado marched north in 1540, his horses were the first the natives had ever seen. But in due course adventurous braves began capturing escaped horses or trading for animals with Spanish colonists. The Plains Apache were mounted by the 1640s, the Kiowa by the 1680s, and the Comanche by the early 1700s.

The conquistador never conquered the plains. "Who could believe," wrote Castañeda, the chronicler of Coronado's expedition, "that 1000 horses and 500 of our cows and more than 5000 rams and ewes and more than 1500 friendly Indians and servants . . . would leave no more trace . . . than if nothing had been there—nothing—so that it was necessary to make piles of bones and cow dung, so that the rear guard could follow the army." Thousands who came after Castañeda and Coronado learned to believe it. But gradually they taught themselves how to live in this mysterious land.

Spaniards also left upon the plains an animal that became the very symbol of Texas—the longhorn, descended in part from the *ganado prieto*, the famous black fighting bulls that Iberians released upon terrified enemies up to Roman times. The wild prairie bull, with his five-foot spread of horns and his absolute lack of fear, was considered by hunters of the day to be more dangerous to a man afoot than the grizzly bear. By the middle of the 19th century Texas longhorns were grazing, free as the buffalo and the antelope, from the Red River Valley on the north to the Rio Grande and beyond in the south, and from the Louisiana line on the east to the Pecos in the west, an area of at least a quarter of a million square miles.

The longhorn was hunted as a wild animal by lancers of the Spanish cavalry, and Jim Bowie is supposed to have knifed bulls from the back of a galloping horse. During the Mexican War a wild bull, maddened by a shot fired by a soldier in a column marching to meet Santa Anna, charged the U. S. Army and scattered a whole column of infantry.

Not until after the Civil War did the longhorn become the herded beast known to every moviegoer. Texas, a part of the defeated Confederacy, was short of cash, and the triumphant North was short of meat. The cowboy turned the longhorn from wild game to domestic beef with a mustang, a rope, a branding iron, and nerve. In the 15 years after 1865, nearly five million head were moved from range to market. The great cattle drives to railheads like Abilene and Dodge City, Kansas, left the bones of many a cowhand lying beneath the bleached skull of a steer. But these cattle, these men, and their epic drives built an empire.

Cattle that can cope with the country are still prized. "Our cattle are *tough*, and that's the way we breed 'em—survival of the fittest," I was told by Chris Lacy. "Lots of our range-fed Herefords don't know there is such a thing as a human being until one man's got 'em by the head and another by the feet."

The Texas longhorn, like the bison that earlier dominated the plains and prairies, is now a rarity, preserved in sanctuaries but also on a growing number of working ranches by those who like to dream by the window of the past. Among the dreamers was one of the supremely practical men of our times, President Lyndon Baines Johnson. On the LBJ Ranch near Stonewall, in the hill country of central Texas, President Johnson had established a small herd of longhorns; and I went there at the invitation of Lady Bird Johnson in the hope of catching a glimpse of them.

With Mrs. Johnson at the wheel of her car, we took a long drive over the meandering roads of the LBJ Ranch. "Lyndon just threw himself into a love affair with this little piece of the world," said Mrs. Johnson. Here the 36th President had built a dam, there a dirt tank to catch the grudging rainfall, here a pen or a fence. He had uprooted every juniper—an act of good husbandry, for this hardy evergreen takes moisture needed by range browse and grasses, and few creatures of the plains will feed upon it. But he had left every oak standing amid broad fields of native grass. Mrs. Johnson remarked that this section of the ranch resembled an English park, and indeed the composition of shade and greensward and fluffy cloud (low enough, it seemed, to be touched by a horseman's hand) was very like a Constable landscape brought to life.

Mrs. Johnson put the other colors into this *(Continued on page 68)*

"Everybody's somebody in Luckenbach," decrees the slogan of this dusty hamlet about 60 miles west of Austin and north of San Antonio. Country-and-western singers Waylon Jennings and Willie Nelson popularized isolated Luckenbach— population 7, more or less. In the "chat room" of the general store (right), visitors gather to gossip, sip cold beer, and relax to the lively music of a three-piece band. At the Luckenbach World's Fair, held 12 miles north at Fredericksburg, fans cheer on such distinctive events as the Intergalactic Chicken Flying Contest. The 1978 winner, Black Diamond (left), set a world record by flapping 254 feet 11 inches. In a more down-to-earth contest (below), the crack of the starting gun sends nine-banded armadillos hurrying toward the finish line. *"There's big money in racing armadillos,"* declares a veteran handler. *"I've won better than $3 in a single afternoon."*

Dawn assault: Smoke screen shrouds an M-60 tank during a training attack across Cowhouse Creek at Fort Hood, Texas. Clanking over an aluminum "ribbon bridge," the 55-ton tank joins other elements of the First Cavalry Division in the river-crossing exercise, one of the most vulnerable of troop operations. At right, standing water from heavy spring rains reflects the flash of night-firing tank guns at the 339-square-mile armor training center, home of the largest concentration of combat forces in the U. S. Army. Military installations have dotted the open spaces of the Southwest since the 1850s.

painting. All over the ranch, along its roadsides and in its upland spaces, she has sown the wild flowers of Texas. We alighted from the car and strolled among them, and Mrs. Johnson told me how each comes in its season: bluebonnet and winecup, daisy and Indian paintbrush, thistle and blooming cactus. "You can't take a footstep without crushing a dozen blossoms," she said sadly as our shoes pressed petals into perfume. "It's like a Persian carpet!"

We never did spot the LBJ longhorns; but we did see even more exotic beasts. Among throngs of native white-tailed deer, animals from other continents roamed: black buck antelope, fallow deer, mouflon sheep. We looked in vain for the axis deer from Asia, a noble animal with a rack of antlers sweeping back over rippling withers. But like the longhorn, the axis stayed hidden in the brush in daylight hours, and would emerge only after the fall of dusk to drink and feed.

That the axis deer should have eluded us came as no surprise to me. A day or two before my visit to the LBJ Ranch, I had accompanied a Fredericksburg lawyer named Joe D. Clayton in an attempt to stalk one of these wily animals in the broken country of Keith Meadow's Roaring Rock Ranch near Mountain Home, Texas. With our noses in the wind, Joe and Keith and I moved for an entire sun-shot morning over rocky stream beds and through thick cover of juniper and oak. We saw sign of the axis—bits of hair snagged on the brush, a track or two, a trace of droppings. But the deer itself, for which Joe would pay $750 if he happened to bag one, was as invisible as a myth.

I knew the axis existed, for Thompson B. Temple, the young owner of the TBT Ranch at Mountain Home, had shown me several splendid examples. The TBT is a breeding ranch; there is no hunting there. Those who seek trophies on the 120,000 acres of hill country controlled by a ranchers' organization called Texotic Wildlife, Inc., must get them in a fair chase, as Joe Clayton had attempted to do.

Some hunters, according to Temple, spend as much as $15,000 a year hunting game from Asia Minor and the Far East in sections of Texas brush country too poor to support profitable cattle ranching. But exotic game operations have a purpose beyond commercial sport. Thompson Temple, by crossing the desert bighorn sheep, which had nearly vanished from the Southwest, with the mouflon from Corsica and Sardinia, has produced a strain that is seven-eighths bighorn and indistinguishable in habit and appearance from the indigenous species. The black buck antelope is a threatened species in India, but it thrives in Texas.

As to the hunting ethic, perhaps a hunter like Joe Clayton, who has bagged many a record trophy, can explain it best: "Those who don't hunt can't understand, I know," he told me. "But I've stalked with gun and bow, and to me, doing that awakens the very oldest feelings man has had. There is something reverent between man and his quarry, and we have known it ever since we painted the first picture of a hunter, a spear, and an animal on the wall of a cave."

"It's a lot of pressure. When the wheat's ready, everybody wants it cut right away," muses independent harvester Norman D. Wedel. He begins in southern Oklahoma in early June, continues into Kansas, then heads west to Colorado for the barley harvest. In a long day that may begin before sunrise and end at midnight, his three machines and crew of six cut as much as 300 acres of grain.

MEN AND WOMEN think deeply and speak their hearts in this wide country. From the beginning it has inspired verse and song, and it inspires them still. The Southwest is filled with the sound of guitars and young voices, singing of loneliness and loss and distance and betrayal and work. At the Kerrville Folk Festival, I sat among 5,000 people as they cheered under the stars for a group of young musicians, any one of whom might be next year's country-and-western superstar. Backstage

I asked a young tenor in ripped blue jeans and a stained cowboy hat who had just brought the crowd to its feet how he had moved them so. "Sir, I just sing about what happens to me," replied Gary P. Nunn, "and I figure that others will hear what happened to them in my music."

A few days earlier, standing in mud with my hand on the cold, damp flank of an M-60 tank, I had wondered what lay in store for the young people who were on maneuvers at Fort Hood, Texas, with the First Battalion, 67th Armor, 2nd Armored Division. I had been shown to the place where the battalion had been bivouacked in driving rain (ten inches in three days) by a whippet-slim, ramrod-straight young second lieutenant in starched fatigues, steel helmet, and pack harness. Her name was Regina Largent, and she had just completed officer's training after a two-year stint of enlisted service.

She handed me over to a company commander, First Lt. Joe Lineberger, who took me on a wild, jouncing ride in his jeep. Air Force jets screamed and Army helicopters clattered overhead. Under camouflage netting, I spoke to bright young tankers and gunners operating old equipment—the M-60 tank came into service 20 years ago—and found them remarkably cheerful.

"Five enemy to one American is normal, sir," a red-headed, 22-year-old tank commander told me. "Ten to one could be the actual odds if war comes. But those folks on the other side will know we're there." Added Lieutenant Lineberger, a cigar-chewing, athletic young blond who had been a Green Beret and a Ranger before becoming a tanker: "If they do attack, we'll spoil their whole day!"

But as more senior officers told me privately, the difficulty of maintaining outmoded equipment and the problem of keeping trained men in an all-volunteer Army pose real threats to the safety of the troops should they be sent into combat. "The 2nd Armored Division is still 'Hell on Wheels,'" one retired general told me. "But it needs new wheels. We're hoping that the new XM-1 tank, due to come into service this year, will give us a machine to match the enemy's."

The Southwest has always been an important territory militarily— first as an outpost for Spaniard, Mexican, Texan, and American, and later as a major training ground for U. S. forces. There's plenty of room to march and countermarch, but there are other reasons for stationing troops here. The Southwest is an intensely American place, proud of its history and believing in what southwesterners still call simply "American know-how."

Peter T. Flawn, president of the University of Texas at Austin, defined this attitude vividly:

"The door to the future is wide open. And it's tremendously stimulating to live in a place that's building and growing. Texas and the Southwest know *what* they are, and the people know *who* they are. Here, there's a sense that what is real is produced from the earth."

What music to an American ear! And I heard it everywhere. At Kerrville's Arts and Crafts Fair, a young woman showed me the exquisite scrimshaw she had etched on the ivory of a woolly mammoth and said, "I feel, as I work, such love for what this earth has given us and has still to give!" Her name is Cathy Holt, of Garland, Texas, and she calls her work "Tracks in Time." And in San Angelo, center of an area that produces a million head of sheep for market annually, I attended the nation's largest sheep auction; a bearded lead goat, aloof as a 19th-century butler and oblivious to the chant of the auctioneer, led the *(Continued on page 78)*

(Continued on page 78)

FOLLOWING PAGES:
Lowering sky portends a late afternoon storm on Oklahoma's southwestern plains. Spring rains and warming sun will ripen this emerald winter wheat into June's golden harvest. Wheat fields stretch for scores of miles across the level lands of Oklahoma and Texas, yielding about 325 million bushels each year. The climate allows farmers and ranchers to graze cattle on the young wheat from November to March, and still harvest a full crop in early summer.

Timesaving cutting patterns mark wheat fields near Pampa, Texas, where combines reap the ripened crop. Planted in the fall, the grain goes dormant through the winter, then grows rapidly when temperatures rise. The truck in the foreground will carry cut wheat to steel storage bins. "Wheaties"—the combine crews—race to escape the threats of wind, hail, and prairie fire.

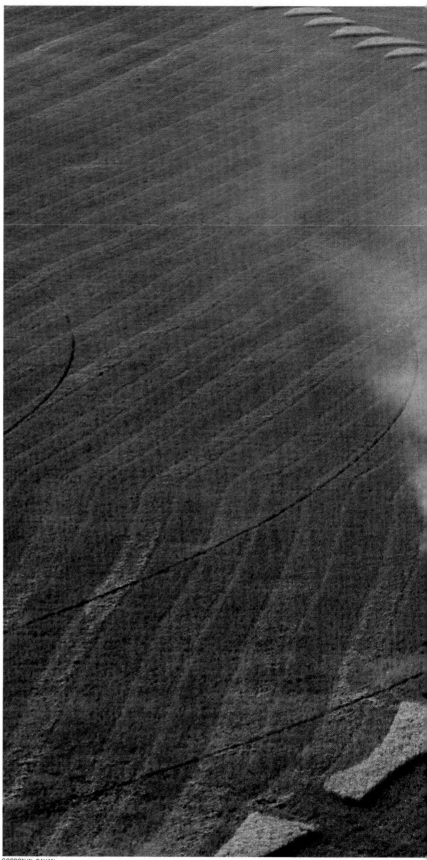

Against a fiery evening sky, smoke billows from the diesel engine of a Baker & Taylor drilling rig near Cheyenne, Oklahoma. A 140-foot derrick supports the gear that struck gas at a depth of 15,500 feet. Thirty miles to the west near Wheeler, Texas, roughnecks make a "round trip"—pulling out thousands of feet of pipe to change a drill bit (right), then running it back. Stacks of pipe sufficient to reach four miles into the ground rise from the drilling platform (far right). A 20,000-foot hole may cost as much as five million dollars—and yield nothing. But the potential profits keep large exploration companies as well as independent wildcatters in the high-stakes competition. Together Oklahoma and Texas produce more than a billion barrels of oil and eight trillion cubic feet of natural gas a year.

Setting sun illumines granite heights of the Quartz Mountains in
southwestern Oklahoma. Below on Lake Altus, the Quartz Mountain
Lodge houses 200 talented students during the two-week Oklahoma Summer
Arts Institute. Chosen by audition, the young people—ages 14 to 18—come
for intensive training under renowned teachers. Aspiring ballerinas, lured by
the prospect of a future with a dance company, train long hours at the
barre; one pauses at far left to adjust a slipper. A chance to perform with
a major orchestra inspires musicians such as the French-horn player at left.
Oklahoma has developed or preserved many of its scenic and recreational
resources in such state-owned resorts as Quartz Mountain.

frantic ewes through the pens. Ewe-and-lamb pairs were going at $50; last spring they sold for as much as $75. "The farmer is the only businessman in America who doesn't mark his prices on his merchandise," writer Elmer Kelton told me. "He says, What'll you give me?—I'll take it."

Elmer is the author of a fine novel called *The Time It Never Rained*. That title sums up a lot; if you can't count on the weather, you have to count on yourself. In the Texas Panhandle, I saw hail the size of baseballs putting dents in parked cars. In Lubbock, they tell of sandstorms that turn day to twilight, and of gusting, dust-filled winds that pluck the books from the hands of coeds at Texas Tech, leaving them clutching at campus lampposts to keep from being turned into the world's prettiest tumbleweeds.

Lon Burks of Wichita Falls, Texas, an employee of the U. S. Weather Service, was awakened at home on April 10, 1979, by warning sirens. Tornadoes were coming up from the southwest. Lon got into his bathtub and covered himself with a mattress. "It turned completely dark," he said. "I could hear the leading edge of the tornado tearing houses apart . . . boards snapping, debris hitting the walls, and the loud roar of the winds. Then it got quiet. The tornado had formed an enormous eye, and for a moment I was in the eye. Then the trailing edge of the funnel hit and ripped the roof off the house. I expected to be blown away with the rest of the debris."

All of Lon's house was destroyed except for his bathroom refuge. He counts himself lucky: In the three tornadoes that struck the area that day, 56 people were killed and 63 million dollars in property damage was reported. Each storm lasted more than an hour and generated wind velocities as high as 350 miles an hour.

IN SOUTHERN OKLAHOMA, as great red combines with 24-foot headers harvested some of the first of the plains' winter wheat, a farmer named Woodrow L. Bohannon told me about the satisfaction drawn from 50 challenging years of working the soil. "My dad homesteaded this country, and he gave me farms so poor that he couldn't rent them to anyone else. The soil had been beaten to death, and I spent ten years before I saw a 13-bushel acre. Since then we've averaged 25 bushels around here, and we hope for 32 to 35. I'm content. I've seen the land come back from ruin."

As the grain fell like burnished gold under a sun as hot as Egypt's, Woody Bohannon told me of battles with an aphid called the greenbug that usually comes in February on the prevailing southern winds. "There may be thousands of them on a single wheat plant, and they inject a toxic substance that kills the plant. A ladybug beetle can eat hundreds of them in a day, but the ladybugs don't reach southern Oklahoma until mid-March."

In recent decades some Oklahoma and Texas growers, faced with a falling wheat market, have turned to growing cotton in the good clay and sandy loam that averages 300 pounds—three-fifths of a bale—per acre. Research has finally overcome cotton's nemesis, the boll weevil, by outwitting it.

The pest larva emerges in the plains about June 20 and can sustain itself by living in the "square" on the young cotton plant that is the embryo of the boll. Without the square, the larva dies in ten days. Now farmers simply wait out the weevil; they plant their fields after it has died. "We fought the weevil for a hundred years with arsenic and Paris

"Yellow Rose of Texas" adorns a $400 pair of brushed-calf boots custom-made by Charlie Dunn. Meticulous craftsmanship and precise fit characterize the work of this famed bootmaker, who began his apprenticeship at the age of 7. In his shop in Austin, Dunn and five assistants produce handsewn boots that command prices as high as $2,400. Opposite, another talented Texan, Kammie Milligan, and her horse Top Hat perform at the Pin Oak Charity Horse Show in Houston.

BOTH BY GORDON W. GAHAN

green, and five years ago we thought we were going to lose to him for good," said Woody Bohannon. "And all we had to do was disturb his life cycle." Of the 14 million bales of cotton grown annually in the United States, nearly half is now produced in Texas and Oklahoma.

Most of the winter wheat grown on the plains is harvested by itinerant crews. In many cases the crew's nucleus is a family, like that of Arnie Balzer of Inman, Kansas, who went into the business when the combines (average price: $72,000) and other heavy equipment became too expensive for individual farmers.

"We start here in Oklahoma in late May or early June with the wheat," Arnie told me with the touch of accent he retains from his German parents, "then we go to Kansas. After that we cut silage around Hereford, Texas. Then we cut high-moisture corn back in Kansas, and wind up on Thanksgiving."

Four big combines moved through the fields as he spoke. "I'm 66 now. My son, Loren, runs the business; his wife, Jane, cooks; my son-in-law is with us. Martha—my wife—didn't come this year, but she was with me every other year since we started in 1946."

"You know, Arnie, that we're expecting you folks Sunday at First Baptist," said Woody.

"We'll be there, because we don't cut on Sunday forenoon," Arnie Balzer replied. "We like to have our boys go to church, and we don't hire them if they drink or smoke. We've always had good boys, and their parents feel they're safe."

Arnie looked to one far horizon and then another. "Wherever the Lord is, that's home," said he. "When this harvest is over, Martha and I are going on a trip to Russia. Our Mennonite grandparents left there in 1874. But I expect their religion stuck it out. Their wheat, too."

Sticking it out. Truman Tucker's folks came into the Oklahoma Panhandle in 1905, and his mother frosted her feet walking beside a covered wagon. She died when Truman was only 6.

Truman remembers the Dust Bowl storms during the long drought of the 1930s. "They went to getting a few tractors out on the flat and plowed it all up. Then the wind started to blow, and it blew most of the time. You'd see dust devils two, three hundred feet high; they'd just hang there and spin in the sky. Lots of days you couldn't see these hills, half a mile away. You never knew where that dust came from—Colorado? Kansas? One day, it was the 12th of April, we were having a funeral, burying a preacher. We looked up and saw that thing coming. We didn't have the grave filled, but we got inside a building. The sun was shining as bright as it is now. The storm turned the light out—just like blowing a lamp out in a room at night."

Truman showed me some photographs. "But most folks stuck," he said. "Some went away; but when the rain started again, they came back. It took a lot of rain and a lot of years. Folks said the grass was dead and would never come back, but the grama grass did, and the buffalo grass. The Lord made that grass for this purpose."

And the people. I stood on a hilltop with Mrs. Lyndon Johnson while the hot day cooled. We could see grazing cattle and straight strong fences and, a bit farther away, the tip of a white steeple. "The country's changed," said that great lady. "It's so much more prosperous now and vigorous, and yet it's still got some of the dear old lovable ways. It's like a boy of 18, 'a-growin' and a-doin'.' I feel it in my bones, and it's a spirit-raising thing that makes the blood run."

FOLLOWING PAGES:
Solitary sentinel of the rural landscape, a windmill breaks the flat horizon of a Texas Panhandle dawn. The wind blows almost continuously across the endless plains, turning windmill vanes to pump water for the region's farms and ranches.

79

The Cities

Spreading and sprawling and reaching skyward, the cities of the Southwest rank among the fastest growing urban areas in the world. Located at a low pass in the Rocky Mountain chain, the border town of El Paso, Texas (above), blends a spicy heritage of Spanish, Indian, Mexican, and the American West into a lively and cosmopolitan city. The Austin campus of the University of Texas (far left), still known as "the Forty Acres," now covers more than 300 acres in the heart of the state capital. The domed observation deck (left) of the 50-story Reunion Tower offers an expansive view of downtown Dallas. Once a frontier trading post, Dallas has become the center of a metropolis of more than two million people.

Designed for retirement living, Sun City, Arizona, clusters moderately priced homes around neighborhood shops, churches, and recreational facilities that include 11 golf courses. In 20 years Sun City has grown to a population of 48,000. Year-round sunshine, mild winter temperatures, and a favorable business climate attract people of all ages to Arizona. At right, while most of Tucson glimmers in a fading dusk, stadium lights blaze at the University of Arizona. In Phoenix, the burgeoning capital, the desert sky reflects in the soaring glass walls of the 40-story Valley Center (far right).

Pushing a curtain of clouds, a "blue norther" rolls over downtown
Houston. The skyline changes almost daily as new structures, many
of them headquarters for major corporations, sprout from the Texas
flatlands. Fueled by a steady influx of people and money, oil-rich Houston
continues to head the list of rapidly growing American cities: Every month
86 more than 4,000 newcomers move into the metropolitan area.

Winding through downtown San Antonio, the oldest city in
Texas, the tranquil, tree-shaded San Antonio River (right) has
inspired a commercial renaissance along its banks. Above, right,
cascading fountains surround visitors to the Water Garden in
Fort Worth—a former "cow town" now building a reputation as
one of the leading cultural centers of the booming state.

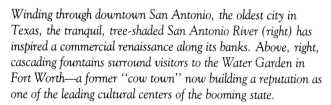

Turned to gold by autumn's Midas touch, aspens
shimmer against the deep green of spruces and the
somber gray of granite in the San Juan Mountains of
southwestern Colorado. The crest of 14,150-foot
Mount Sneffels overlooks a highland wilderness rich
88   in history, in mineral resources, and in scenic beauty.

"The shining temple and fount of the Southwest...."

THE MOUNTAINS

IOPENED MY EYES on the Fourth of July and saw the flag of the dawn—a patch of red sunlight reflected on a field of snow—climbing toward the morning star, which shone for a last moment between the black of night and the blue of day.

If the colors were right, so was the music: the rustle of aspen leaves in a gentle breeze, the rush of a mountain stream in the place where I had made my camp, 12,000 feet above the sea in the San Juan Mountains of Colorado.

I was alone, but of course I was not alone. This was a natural camp, a soft place where a little grass grew on the stony face of the mountain—a place where there was new water, sun-melted the day before, a place where a ledge and a grove of trees provided shelter from winds that still blew bitter cold when the sun was down. Others had camped here: the Ute and those who came before the Ute; the trapper and the hunter, the miner. Their fire circles and their ghostly footprints were all around me.

In my imagination, I saw an American with a feather in his headband, broiling a trout here on July 4, 1776. What would he have replied if, somehow, I could have stepped across 204 years from my morning on the mountain to his and told him what was occurring 1,800 miles to the east in a muggy city called Philadelphia? There, a group of rebellious subjects of a king who resided still another 3,500 miles to the east were solemnly voting to adopt a Declaration of Independence—thereby transforming themselves into Americans and their simple agrarian colonies into a union whose citizens would, in only eight generations, leap from the plowshare to the spaceship. No doubt my original American, on hearing that, would have given me his breakfast, for The People were ever generous and good to madmen.

There had been madness in these mountains. As sunlight tugged at shadow, I caught a glimpse across the valley of the buildings of an abandoned mine. Weatherbeaten, leaning drunkenly, they clung to the hillside as a prospector clings to his dream of gold. Here, abundant gold and silver were realities in the last three decades of the 19th century.

Although the boom ended long ago, mining continues in the San Juans, and living men still remember the afterglow of the glory days. In Durango, which became a smelting center for the mining region, I heard Arthur Isgar speak of the human cost. "My father died in 1936 at the Gold King Mine," said Art. "He'd been a miner all his life, and he worked all that day and died in his sleep, with two of his sons near him in the same bunkhouse." All four of Art's full brothers were miners a good part of their lives. He chose ranching as a career, and he is the only one of the five still living.

The Spaniards prospected in these mountains, and local legend holds that the Ute and other upland tribes knew of the existence of gold but kept the secret for fear of losing their hunting ground to feverish whites. Mountain men, who came in search of beaver pelts to be made into tall hats for European and American dandies, were hardly more talkative than the Indians, and for the same reasons.

The Spanish law forbade foreigners to carry on trade within the empire of Spain. Trappers risked the fate of one expedition whose furs, valued at $30,380.74½, "the fruits of two years' labor and perils," were confiscated in 1817 by the Governor of New Mexico. But if the authorities in Santa Fe were strict, mountain men weary of isolation found a season in hospitable Taos a highly congenial experience. This adobe town lying at the southern end of the mountains was, as the popular historian Harvey Fergusson wrote, "where corn grew and women lived. . . . More than any other place . . . the heart of the mountains."

During the long period of Spanish dominion, a visit to Taos could be a thrilling business—especially if Utes, Apaches, and Comanches all happened to be in the plaza at the same time, as was sometimes the case, and if they happened to be at war with one another. There was a brisk

Ridged by the Continental Divide, their snowfields the source of the Southwest's major rivers, the southern Rockies encompass arid canyons, evergreen forests, and snowclad peaks. Here mining and ranching still provide a livelihood for some, though increasingly the land feels the impact of recreational and residential use.

trade in everything, including human beings. The fiercer Indians sold to the Spaniards, and sometimes to other tribes, the captives they had taken the year before. The practice continued under Mexican and then United States rule: A young, healthy Indian could be bought for $300 even after New Mexico became U. S. territory.

That memory brought a rueful smile to the lips of Fray Angelico Chavez, a Franciscan historian and poet who is descended from one of the first families of New Mexico. "People here boast of the pure Spanish blood that runs in their veins," he told me with a twinkle of scholarly mischief. "But perhaps they get some of their pride from the blood of the Kiowa and Pawnee that were bought from the Comanche and taken as wives—and from those wild Anglo-Saxons who married our women later on."

But on that Fourth of July morning in the San Juan Mountains, I was living, however briefly, the life that men long lived among the peaks—a life alone with nature. Some live it still. In the La Sal Mountains of Utah, I had spent time in a mountain cabin with an ageless cowboy named Chet Smith. Texas-born and Colorado-raised, Chet had been in the country around the La Sals since 1927, and had been working for the family of rancher Hardy Redd for more than 40 years. "I can't live in a town," said Chet, warming up a pot of stew for Hardy and me. "I've got to be out where I can't hear no noise." Chet paused. "Hardy, what's that blankety-blank *hum?*" That "hum" was a machine working underground in a mine ten miles distant. "Too close," said Chet, and served the stew.

The mountain men lived on into Chet Smith's time. He remembers a trapper named Burt Rowel who lived high in the La Sals. The snows can come in October, and they can be deep; there's no getting out once they've fallen. "Burt would kill a porcupine when he saw one and hang it up in a tree," Chet recalled. "If he got *(Continued on page 97)* 91

Pueblo Indian women await buyers for silver
and turquoise jewelry and other handicrafts
beneath the restored portal of the Palace of the
Governors in Santa Fe. Banners on posts
herald the city's traditional Fiesta, a
celebration observed every summer since

1712. In August, Santa Fe holds its annual Indian Market, where skilled artisans like silversmith Joe H. Quintana, left, exhibit their work. Except for one interval the palace—oldest public building in the United States—housed Spanish, Mexican, and U. S. governors of New Mexico from 1610 until 1909, when it became a museum. Indians controlled the palace for 12 years after the bloody Pueblo Revolt of 1680 put Spanish settlers to flight. 93

Snaking ribbons of light from handheld candles mark the course of a solemn procession to the Cross of the Martyrs, climax of the Santa Fe Fiesta. The cross memorializes the Franciscan victims of the 1680 revolt, when Indians of all the pueblos rose in protest against Spanish mistreatment. After the procession the city celebrates the reclaiming of New Mexico in 1692 by Diego de Vargas.

caught in a storm he was all set—he'd find him one of them froze-up porcupines, cook it, and sleep by the fire under the tree where he'd hung it."

Gone, though, are the days when Chet was a top hand on the Redd ranches and each spring the cowboys would trail hundreds of Herefords through the Paradox Valley to summer pasture in the high country.

"It was a pleasant thing, a cattle drive," Hardy told me. "You'd start early, push along at a steady walk for three or four hours, then let the cattle drift along, grazing. Around noon you might stop for shade from the heat. Finally another couple hours trailing in cool evening; then a campfire and supper, maybe venison, and you'd eat all you could hold. We truck the cattle now. But the pace of a horse following a cow, to me that's more man's natural speed."

To SOME WHO GLORY in the mountain winters today, the natural pace of man is that of a speeding bullet. They are downhill skiers, as special a breed in their way as any mountain man ever was. They may be ragged except for their big expensive boots and their skis, but they are obviously rich in the mystery and the rewards of their lives.

At Purgatory, a ski resort just north of Durango, I tried to penetrate the skier's mystique in a conversation with a young ex-Vermonter named Dirty Don Hinkley. Dirty Don *lives* for skiing. A college graduate, he works in winter as a ski patrolman and in summer as a lumberjack, cutting ski trails. On those small earnings he manages to support a wife, a small son named Ethan, and a pack of Malamutes. Sometimes, using a .50-caliber muzzle-loading Hawkins rifle, he brings home an elk to their log cabin to supplement the larder.

Dirty Don, a former alpine ski-jumping champion, talked excitedly of the upcoming North American-World Speed Skiing Championships scheduled for April above Silverton. "The world record is 124.04 miles an hour," he said. "But the computer prediction is that 138 is attainable under optimum conditions." At 30, Dirty Don figured he was still young enough to try for that record.

How long would he go on skiing? "Forever," he answered happily, as new snow fell on the steep mountainside outside the window. "A friend kept on as a ski patrolman until he was 63." Why? Don gave me a tolerant look. "It's a spiritual thing," said he, shouting to make himself heard above the din of disco music that filled the restaurant of the ski lodge where we were sharing a cup of hot chocolate. "When I ski, alone above 10,000 feet, I'm part of the snow, part of the cold, part of the mountain—100 percent in a natural element."

Thousands of young men and women answer the call of these mountains. Durango, a town of 12,000 lying a few miles north of the New Mexico line and a few miles east of the boundary with Utah, has become a mecca for them. "The word went out, Lord knows how, that Durango's the place," an old-timer told me. "They liven things up, I'll say that." And so they do—Ph.D.'s pumping gas, M.A.'s waiting on tables. They stay long enough to fall in or out of love, fly or fall down the mountains on skis or hiking boots, according to their skills—and then vanish.

Durango, a mining town that has died and risen from the ashes of more than a few booms, glitters now with reproductions of miners' saloons, facsimiles of dancing girls, and even a genuine narrow-gauge steam train that makes a daily run in summer to Silverton, 45 miles to the north. Once again it's gold of a sort that has brought the rush of newcomers—the treasure of solitude and majestic beauty; the last chance,

Six-century-old ties bind tradition-conscious Taos Indians to their clustered multistoried adobe village. Accepting change only slowly, the people of Taos Pueblo permit no electricity or plumbing in parts of the old town. Formerly residents entered their houses through the roofs, but now most houses have doors and windows on the ground floor. Unlike other southwestern pueblos, usually built around a single plaza, Taos comprises two separate complexes on opposite sides of a stream. On the north side stands the Roman Catholic Mission of San Geronimo.

perhaps, for the urban-bred to be "100 percent in a natural element."

Ironically, efforts to preserve this natural element may be having the opposite effect. The San Juan National Forest encompasses 2,086,484 acres, much of which, under the multiple-use policy of the Forest Service, is lumbered, grazed, mined, designated as protected watershed or wildlife habitat, or reserved for human recreation. Old mining roads and other trails offer access.

In the Wilderness Act of 1964, Congress provided that certain lands in national parks and forests might be set aside "in their natural condition . . . to secure for . . . present and future generations the benefits of an enduring resource of wilderness." Such wilderness areas are closed to motor vehicles and most commercial enterprises, and no structure may be built within them. Mining claims may be filed until December 31, 1983; after that, though the working of established claims may continue, no further exploration for minerals is authorized.

Something more than 400,000 acres of the San Juan National Forest has been declared a wilderness area. "A wilderness . . . is hereby recognized as an area where the earth and its community of life are untrammeled by man," reads the Wilderness Act. In 1965, the first year after the passage of the law, 44,700 visitor-days were recorded in what is now the section of Weminuche Wilderness west of the Continental Divide. By 1979, that number had quadrupled to 187,700.

The wilderness, in short, is not exactly untrammeled by man. "I've been drinking from mountain streams all my life," a veteran forest ranger told me. "But now I'm mighty careful that I know what's upstream." In some wilderness areas in the West, the authorities have "rationed" visitor-days to prevent destruction of the environment by those who would, as it were, love it to death. But in the Weminuche Wilderness, no permit is required to enter the area. "In effect, we're running a regulationless wilderness," said forester Neil Edstrom. "Many of our visitors practice 'no-trace camping'—leaving no sign of themselves behind. Others, of course, are not so careful, and we haven't the rangers to police so vast and wild an area.

"The job of the Forest Service is to preserve wilderness, but also to provide recreation." Neil gave a puzzled shake of the head. "No one has told us how to make those two missions compatible."

ARTHUR ISGAR, born in these mountains, has seen a good many changes in his 65 years. Most of them, he thinks, are for the better. "We used to drink out of the same tank as the horses and cattle," Art told me. "Lots of days you could hardly see Durango from the mountains for the coal smoke, and a housewife couldn't hang out her laundry because of the soot. I'll take today!"

When Art and his wife, Anne, built their first ranch house, they planted trees before they laid the foundation. "You've got to leave the land better than you found it," said Art. He pointed to a tree he had planted. "I watch that aspen grow like a person. I dread the day when I won't be able, physically, to get up in the morning and irrigate. To make things grow is the basic love of a farmer."

The Isgar family farms and ranches a large spread, put together acre by acre over a lifetime of toil, some of it on the "dry side" south and west of Durango where there's only partial irrigation. While Art and Anne and I lunched on the sun deck of their house, the Animas River, at flood stage a few hundred yards away, filled our ears with its boisterous song.

"The space and quality of light in the Southwest" inspire Nancy Kozikowski, above, and her husband, Janusz. Nancy spins wool at her wheel; Janusz, opposite, winds a bobbin beside a two-harness loom that holds an unfinished tapestry. Contemporary craftsmen like the Kozikowskis, many of them recent arrivals in the Southwest, have added a new dimension to a crafts scene formerly confined to traditional Indian and Hispanic works.

Yet Art was in a hurry, after irrigating all morning, to get back to work before his crops burned up. The thermometer stood at 95°F. He told me what drought could mean to a farmer from year to year. "Four years ago, a dry year, we made 187 bales of hay on the entire ranch. Each of the last two years, with rain and some irrigation from snowmelt, we've made around 25,000 bales on the same land, plus 14,000 to 15,000 bushels of grain. We need three things to work with—sun, soil, and water. We have two. Shall we use the third?"

If a plan being studied by the Interior Department's Water and Power Resources Service and a citizens' advisory group is approved by area residents, the answer to Art Isgar's question will be "yes." The Animas-La Plata Project would divert a portion of the Animas River just below Durango and pump the water uphill into a reservoir. The project would provide municipal and industrial water for Durango, several towns in New Mexico, and a portion of the Navajo Indian Reservation in New Mexico. It would also supply 2,000 people living on 70,000 acres on the dry side of Durango, as well as the Ute Mountain Utes and the Southern Utes, who hope to develop the 400 million tons of coal that are thought to lie under their land.

The project would cost 406 million dollars at January 1980 prices. All construction funds advanced by the federal government would be re-paid by those benefited. Water that now flows into the San Juan River and thence into the Colorado would pause to make things grow near its source, in the Colorado and New Mexico mountains. How great is that source? Says John Brown, projects manager at Durango for the Water and Power Resources Service, "Peak runoff from one day's snowmelt down the Animas, 16,000 acre-feet, would service all of Durango's needs for a full year."

There is opposition to the project. Some fear it would displace wildlife, cause unseemly growth in Durango, perhaps attract polluting industries. "Some would prefer that the Southern Utes not develop their coal," said John Brown. "But the Utes have first rights to this water; and they have assured their neighbors that they don't intend to build smoke-belching power plants. If the people want this project, and Congress ap-propriates the funds, we'll build it for them."

It's possible that construction could begin in 1981, and that water could come to Art Isgar's fields and to the Utes ten years later.

These mountains, the shining temple and fount of the Southwest, have ever beckoned outsiders. To the Anasazi who built the great cliff dwellings at Mesa Verde, between Cortez and Durango, they were an Egypt of fertility and civilization. To the Ute, they were a fortress teem-ing with game, from which raiding parties could sweep down like Mac-edonians upon their prey in the villages below—first the Pueblo Indians, then the Spaniards, finally the Anglos. To North Americans, as we have seen, the mountains began as a place to plunder and have ended as a place to find the peace and satisfaction that money cannot buy.

Artists have been captured by these mountains from the earliest days. In the remote village of Medanales, New Mexico, a couple of young weavers named Janusz and Nancy Kozikowski filled an afternoon with nimble conversation and the bright hues of their remarkable work. Janusz, born in Poland weeks before the German blitzkrieg of September 1939, and Nancy, a native New Mexican, met and married in Albu-querque when both were students. They lived in Oregon for a time, but New Mexico called them back eight years ago. *(Continued on page 108)*   99

Treasures of New Mexico: painter Georgia O'Keeffe and her vision of "Black Place III," a stark mountainscape where she camped one stormy night. Born in Wisconsin and educated mostly in the East, Miss O'Keeffe has long expressed in words and in paintings her attraction to her adopted state's mountains and deserts. She bought her Ghost Ranch house in 1940 and a place at Abiquiu a few years later, becoming New Mexico's best-known artist-in-residence. At 93, her enthusiasm for life and nature retains the vigor of her youth and reflects the strength with which she has endowed her remarkable work.

With a new lease on life, retired workhorses of the mountains give visitors a grand view of high ranching country. The Cumbres and Toltec Scenic Railroad (at top), owned by the States of Colorado and New Mexico, runs its steam-powered trains between Antonito, Colorado, and Chama, New Mexico, over a 10,022-foot pass. Richard Braden, above, a ten-year railroad veteran, tends one of the line's 13 locomotives. At right, the Silverton Train of the Denver and Rio Grande Western steams in the yard at Silverton, Colorado, before the run to Durango.

*Rocky pinnacles
of Colorado's Ophir
Needles overlook
a mountainside ablaze
with aspen gold. Prickled
gooseberries, above,
glisten in the sun near
Ridgway, Colorado.
Opposite, purple daisies,
yellow sunflowers,
bluebells, red Indian
paintbrush, and white
parsnip make a garden
of a meadow below
Stony Mountain near
Telluride, Colorado.*

In an idyllic setting in the San Juan Mountains, a working ranch survives amid stirrings of the winds of change. Gary and Pat Everett manage the Valley View Ranch, a 2,300-acre spread where they run a herd of 650 cows and yearlings and also raise quarter horses, the Southwest's most popular breed for short-distance racing as well as for cow-pony and general saddle-horse use. But this mountain sanctuary near Pagosa Springs may soon fall victim to its own allure as it attracts newcomers who seek solitude amid the scenery and developers eager to subdivide the land into vacation retreats.

Using natural fibers and natural dyes, the Kozikowskis produce tapestries that have won them an international reputation. Nancy lifted the lid of a boiling pot where yarn bubbled among onion skins: "The longer it cooks, the darker the color; you take it out when your eye tells you."

Nancy gave one of her tapestries to Pope John Paul II. It is a piece of many colors in which Poland's famed Black Madonna of Jasna Góra

At summer's end, sheep leave exposed high pastures for more sheltered winter ranges. Jim Cooper of Montrose, Colorado, and his dog drive a flock along a highway running past Trout Lake. A sheepman all his life, Cooper maintains a herd of some 1,800 ewes.

seems to be thinking the thoughts of a Pueblo maiden. It is now in the Vatican Museum. Janusz showed me a series of "ghost rugs" he is weaving for an exhibition. The dominant color was blue, and Janusz spoke of the New Mexico sky to explain why this hue had seized him in his work. His wife looked on through eyes bluer than any I had ever seen, and I reflected that if Janusz looked into them long enough he might capture the very azure he was searching for.

Taking my leave of Janusz and Nancy, I entered a lonely region where men seek to experience the agony of Christ. In the uplands north of Santa Fe live the Penitentes, a society of passionate Catholics. During Holy Week, ritualistically flagellating themselves with whips of woven yucca, carrying enormous wooden crosses along the sorrowful way to an imaginary Calvary, they worship their Saviour by reenacting the events of His suffering. No responsible historian credits sensational tales that Penitentes have died lashed to their handhewn crosses, but in earlier times some, in their fervor, remained hanging on the symbol of their faith until they lost consciousness.

No outsider is admitted to most of the rites of the Penitentes, but as I passed through their villages, all dust and silence and strong sunlight, I sensed a different atmosphere. Living in isolation for generations, the people of these mountains seem to have mixed medieval Spanish mysticism with an austerity consistent with their remote and rugged surroundings. Perhaps what I could not quite hear in those murmuring towns was the moody discourse between two ages shackled together.

In Chimayo, I found a cheerier union between imported faith and native earth. In the graceful adobe Santuario de Chimayó, built in 1816, I visited a shrine containing a well of holy soil. Nearby stands a figure, richly dressed, of the Santo Niño de Atocha. The fine dirt, when mixed with water and drunk or smeared upon an afflicted part, has effected miraculous cures for believers.

The little chapel was adorned with pairs of baby shoes. I asked a woman, after she had finished her prayers and gathered a kerchief full of soil, what purpose the shoes served. "Ah," said she in surprised Spanish, as if all the world knew the habits of the doll in his glittering gown. "Santo Niño leaves the church at night and wanders the hills, wearing out his shoes. So those who love him bring him new ones!"

IF AMERICA has an eternal city, it is Santa Fe. Certainly, at 7,000 feet, it is closer to heaven than most. Santa Fe has always reminded me of Rome. The colors are the same—terra cotta in the glare of midday, blush of rose when the sun rises and sets. It has a Roman air about it: a dustiness that is the perfume of a long history, an energy that is the sign of a people so long at home that they stride through the streets with the confidence of an old family moving about the house in which all their generations have been born.

Like much of New Spain, Santa Fe was first settled mainly by those from Extremadura. "Those willing to come were offered the aristocratic title of *hidalgo*," says Fray Angelico Chavez. "What bait! But a relative from Extremadura visited me here, and when she saw the countryside, she cried, 'This *is* Extremadura! This is home!' Chavez, or *chaves*, is the Galician word for 'keys,' and I guess we fitted."

If the first families of New Mexico found here the mirror image of their ancestral land, others have had more trouble adjusting. "This landscape is overwhelming—it smashes sculpture, paints over paintings," explains John Fincher, a young Santa Fe artist whose work is just beginning to win recognition. At 39, John Fincher may have found a way to deal with a subject as immense as the Southwest. He paints enormous pictures of small objects—a boot heel three feet high, a spur the size of a horse's head, a magnified prickly pear.

I needed help to say goodby to these mountains, and on a night in July I turned to an old friend, Wolfgang Amadeus Mozart, to find it. The Santa Fe Opera was staging Mozart's *The Magic Flute,* and the splendid music was never sung by a handsomer cast. Mozart, I think, would have liked the setting as much as I did. It was the very amphitheater of Olympus in which those tenors and sopranos, baritones and basses reverberated. Pools of sunlight lingered like magical lakes upon the mountainsides while spears of lightning split the sky.

Then the first notes of the overture soared into the dusk, and the music touched the mountains. I remembered the peaks as I had seen them at dawn in their whiteness, and the song of their rushing waters seemed a fitting counterpoint to the rippling splendor of Mozart's arias.

110

Prairie sunflowers take a precarious hold in the dry, shifting sands of Colorado's

Great Sand Dunes National Monument. High above the hot dunes, clouds cool the peaks of the Sangre de Cristo Mountains. At left, an ant seeks food in the tiny oasis provided by one of the venturesome plants.

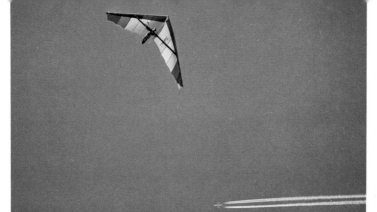

Hang glider and jetliner (left) share the sparkling sky above Colorado during the Telluride Hang Glider Invitational Fly-in. Below, a hang-glider pilot swoops from a mountain crest bound for a landing at the Telluride town park about 5 miles away. One glider flew 30 miles to Silverton, and another made the 25-mile flight to Ridgway. Festivities during the week-long meet included impromptu music-making with an old-fashioned washtub bass, an instrument sometimes called a gutbucket. At bottom left, Brian Jensen jokes after a happy landing. Sixty pilots participated in the September 1979 event, some coming from as far away as Australia and South Africa.

Lights go on at Silverton, Colorado, as the last of the day's sun shines on lofty Storm Peak. Still a working mining town, Silverton doubles as a summer resort, and also played host to the North American-World Speed Skiing Championships in April 1980. At left, a racer hurtles through gates during the Pro Downhill Race at nearby Purgatory. Below, dawn kindles icy crests of the Needles Mountains in the Weminuche Wilderness.

PAGES 116-117: A cross-country skier and his dog leave their exuberant tracks on virgin snow near Molas Pass south of Silverton.

Extending in its incomparable majesty for more than 200 miles, the Grand Canyon cuts an awesome course across the northern Arizona highlands. Here the late afternoon sun falls on Brahma Temple (center), Wotans Throne (upper left), and— on the horizon—Humphreys Peak, near Flagstaff.

"These walls are the great rock pages of the book of creation...."

CANYON COUNTRY

DAYDREAMING beneath a narrow ribbon of indigo sky that twined round the brow of the canyon, I rested my oars for a moment in the swift brown current of Utah's San Juan River. Here, in a silence pierced only by the hallelujahs of the water and the wind that had made this great furrow in the earth, I floated half a mile below the surface of the planet and hundreds of millions of years back in time.

Such reveries are dangerous on the wild rivers of the canyons of the Southwest. I was wakened from mine by the melodious contralto of Marilyn Rivas, one of our guides. "Pull, Charles, *pull!*" she cried, and I put my back into my response. Ashen oars bent as I fought the current to avoid a rock that lay in my path just beneath the roiled surface.

Ahead of us, already bobbing through the rapids, were the other six boats of our little expedition. I heard a whoop of exhilaration from our leader, Millit Gray. The speed of the water increased to something like 12 miles an hour, 4 or 5 miles faster than it was above the rapids. We were flying.

Or swimming. I was sitting in a dozen gallons of San Juan water. We pulled ashore to bail, and I felt a secret satisfaction that other members of the party, even experts, were scooping water out of the seven-foot, bright orange boats of molded plastic that were the wings carrying us through this remote canyon.

There are much more famous canyons in the Southwest than that of the San Juan, but none to my eyes is more beautiful. And none, I think, is a better place to read the history recorded through time on the rocks of this harsh, far country by wind and water and upheaval, by the fin and hoof and claw of speechless creatures, and by the hand of man.

Here forces deep underground buckled the earth and lifted up mountains. Here the beds of ancient seas still carry the scent of their teeming life—the odor of petroleum trapped in shale. These canyon walls are the great rock pages of the book of creation, and it is symbolic that a man standing against them is no larger than a comma.

The San Juan River drains 25,000 square miles as it winds some 360 miles from its headwaters in Colorado's San Juan Mountains to rock-walled Lake Powell. Much of the land through which the river flows is barren or only sparsely vegetated, and the gallant trees and grasses, shrubs and cactuses that do endure seem to grow out of rock that is no more moist than a licked lip on a day of burning sun. When rain falls, it seldom comes gently, but in thunderstorms that can make you imagine the gods are having a game of bowls with the very planets.

We put into the river on a May morning while thunderheads boiled above the cliffs of ruddy sandstone that tower over the old Mormon settlement of Bluff, Utah. We intended to spend six days on the river to complete a voyage of about 85 miles. "Intended" is hardly the word: With the exception of the highway crossing at Mexican Hat, once you are on the river the only way out is by boat, unless you want to climb the cliffs and walk for miles over the slickrock to the nearest drink of water.

I knew what thirsty work that could be, for the year before I had done some walking over the slickrock. In 1879 Bluff was the final destination of a band of Mormon pioneers—250 men, women, and children who, obedient to the command of church elders, left their farms near Cedar City in southwestern Utah and struck out for the San Juan country. To reach it they had to traverse hundreds of miles of unexplored, unwatered country, then cross the Colorado River and the broken uplands that lie to the east of it like a pile of polished bones.

The Mormons had thought that the journey, perhaps 200 miles as a buzzard flies, would be completed in six weeks. It took six months and covered at least 325 tortuous miles, and the legend of it still rings in the San Juan country like last Sunday's church bells.

A century after this great trek, I had retraced the footsteps of the pioneers. With Lynn Lyman, whose uncle Platte D. Lyman was one of

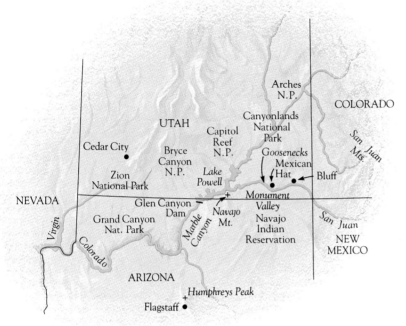

Six parks and nine other preserves of the National Park Service grace the canyon country of northern Arizona and southern Utah. Paradoxically, these sanctuaries contain some of the most beautiful, yet most forbidding and barren land in the nation. Natural laboratories of geology, they provide a graphic two-billion-year record of earth's physical history. Obstinate rivers, volcanic eruptions, violent storms, and relentless winds have sculptured a unique topographical sampler.

the leaders of the expedition, I came in sight at last of the landmark that gives the trail its name: Hole in the Rock. This is a great gash in the western wall of Glen Canyon, 1,200 feet from its top edge to the old riverbed now covered by the waters of Lake Powell. The pioneers built a road descending this near-vertical incline, and down it somehow they maneuvered their wagons.

We had been climbing for several hours, with the morning sun ironing our shirts to our backs, and I paused to catch my breath. Lynn, despite his 73 years, was unbothered by the exertion, and he lifted a hand sculptured by a long life in a hard country to point out the ruts left in the stone by the wagon wheels. "Impossible," I gasped. Lynn smiled gently in reply. His father, as a teenage boy, had driven one of those wagons, and Lynn knew what Americans made of spunk and whipcord and filled with the power of faith could do.

Riding the river current, we knew our way through Hole in the Rock country would be much swifter and easier than that of the pioneers. Trussed up in life jackets, our backs braced against waterproof black rubber bags that held our sleeping bags and spare clothes, we began to learn to handle our agile, unsinkable craft. Millit Gray, a former schoolteacher who gave up the classroom when wild rivers began to run like a fever in her blood, has navigated the San Juan as an expedition leader half a dozen times, and she knows its glories and its pitfalls as well as anyone. It was her job, with the help of Marilyn and a young man named Randy Tate, to teach us how to run a swift river while actually doing it. I hoped that Nancy and I, together with photographer George Mobley and researcher Jennifer Urquhart, would be good students.

Randy, who has been a schoolteacher and a chef, has the far-seeing eyes and the agile movements of his part-Cherokee father and the long blond hair of his Irish-American mother—and a gift of phrase that is all his own. "Just remember the four rules of river running, and you'll be fine," he told us. "Keep your boat in the water; keep the water out of your boat; keep yourself out of the water—and don't mess up!"

To control a boat, Millit told us, you have to go faster or slower than the current. "No human being can fight the river and win," said Millit, "but you can learn to dance with it. Just remember that you're dancing with a mammoth."

The current took us as we rowed away from shore, and we passed under the Sand Island bridge at a brisk 7 miles an hour. The bridge, which links the country of the Mormons to the north with the vast land of the Navajo to the south, was almost the last object made by white men that we would see in the half-dozen days that lay ahead—days so crowded with incident that they would seem to pass in and out of our minds with the fleetness of deer leaping from covert to covert.

For a while the river flowed mildly through bottomland. Millit pointed out movement on the surface, combers smaller than waves but much larger than ripples. "Sand waves," she called, steering us well away from them. The eighth member of our party, a homespun geologist named George Billingsley, told us that sand waves are a phenomenon of fast-running, silt-laden rivers. "Sand is suspended in the moving water," he explained, "and the load increases with the speed and volume of the water. Finally the sand and water reach a harmonic rhythm that sets up the wave effect. Sand waves can be little fellows, like here. Downstream, if the water is high enough, they may be six feet tall. We'll learn to ride 'em. *Fun?* Why, it's better than sliding down a canyon wall on a scared mustang."

Distinctive mesas and buttes of Monument Valley—including the East and West Mittens, at far right—rise more than 900 feet above the valley floor. Familiar to generations of western-movie fans, Monument Valley straddles the Utah-Arizona line on the Navajo Indian Reservation. In the distance rises 10,388-foot Navajo Mountain, one of the legendary heights sacred to the tribe.

HALFWAY TO THE FIRST night's camp, we beached our little flotilla at the mouth of Butler Wash, now a dry bed of polished boulders and coarse sand but a raging torrent in times of thunderstorm. Here once more we saw the hand of man. Spectacular petroglyphs between three and five feet tall, conceivably supernatural beings lent human form, were pecked into the sandstone cliff walls. A short distance farther downstream we stood in what remained of one of the thousands of cliff dwellings of a vanished people, the Anasazi. The stone walls those ancient masons had laid still stood, firm and strong, and the round kiva that was their holiest place was still there, too. I chose not to go inside, so as not to disturb whatever spirits might linger; and when, toying in the dirt while the others explored, I turned up an ear of corn no larger than one of the fingers of the hands painted on the cave wall, I carefully returned it to the parched earth that had kept it safe over the ages.

The Anasazi, a race of builders and farmers living gently on this land, are believed by some to have been the ancestors of today's Pueblo Indians, but much of their story is a mystery. About 800 years ago they abandoned their settlements and dispersed into the mists of legend. No one is sure where they went, or why. From tree-ring research, scientists know that the climate changed; perhaps there was no longer enough rainfall to support their towns. Disease or nomadic invaders may have driven them out. But no one knows for certain, and the Anasazi speak only from the ruins of their unique civilization.

George Billingsley ventured no theory of his own about the puzzle of the Anasazi, but he did have a prophecy about the weather. "It's going

Born in controversy, Lake Powell now wanders peacefully along the Arizona-Utah border, concealing the riverbeds and side canyons flooded by construction of Glen Canyon Dam. Created by the dam's completion in 1963 despite passionate protests by conservationists, the reservoir and its 1,800-mile shoreline form the Glen Canyon National Recreation Area. On the eastern horizon, snow lingers on the heights of Navajo Mountain.

to clear before we hit camp," he told me. Now, I have camped often in the desert, and when an Arizonan—George is from Flagstaff—tells me that it isn't going to rain in the night, I receive the news with a certain amount of skepticism. If it is going to rain on a campsite in the Great American Desert, it's a good bet that it is going to rain at two o'clock in the morning. But, sure enough, the sun broke through the overcast at about five o'clock as we beached our craft on the south shore of the river.

Still wary, I pitched a tent after helping to gather the evening's firewood. There were plenty of cottonwood logs lying about the grassy copse, most of them sharpened to a neat point by the teeth of beaver, whose industry would provide us with fuel throughout the voyage. It was Randy's night to cook, and he turned out a sumptuous three-course supper. His key implements were the traditional tools of the chuckwagon cook—pliers to lift iron Dutch ovens off the glowing coals, and a shovel to manipulate the embers. No better repast could have been produced by a tyrant of cuisine in a tall white chef's cap. Randy operated according to good-humored rules of democracy: Everyone helped chop vegetables, and everyone washed dishes in the steaming pots of river water. The San Juan's water also made delicious coffee—brown as a Navajo cheek, it was the proper color even before the coffee was added, and the grains of river sand mingled with coffee grounds to give the brew just the right texture.

No amount of coffee or campfire yarning could keep us awake. The sun waved a pink scarf in farewell, and Mars, Jupiter, and the star Regulus appeared in the southern sky in a triangular conjunction that would be the night sign of our voyage. Soon afterward, we were in our bedrolls. During the night I woke to the sound of our friends the beavers, slapping their tails on the water as if to frighten our boats away. When the boats neither fought nor fled, the beavers swam on.

WE WERE CAMPED across the river from one of the storied places on the Hole in the Rock Trail, San Juan Hill. This had been the last big obstacle conquered by the Mormon pioneers in the spring of 1880. I had stood here with Hardy Redd, grandson of Lemuel H. Redd, one of the pioneer company who had become a legendary rancher with sheep and cattle roaming over thousands of desert and mountain acres. Hardy Redd, standing beneath a prayer—"We thank Thee, oh God"—carved in a rock, had described this final barrier and the suffering of the horses as they struggled up the slickrock slope. "The steepest places were marked with the blood and hair of the animals' knees where they floundered and fell and yet held their loads," said Hardy. "My grandfather was a strong man and reluctant to display emotion, but he never told that story without tears falling from his eyes."

Our party made the climb, and sitting atop Comb Ridge with our feet hanging over several hundred feet of empty space, we looked round the compass to read the geological history of this land. George Billingsley, standing on the lip of the ridge, showed us De Chelly sandstone of the Permian age (estimated by geologists to be about 240 million years old), Chinle formation (190 to 200 million years old), and Navajo sandstone that was formed a mere 135 million years ago.

To the west, where the river made a great bend, rose a jumbled formation called the Mule Ear diatreme because of its outline as seen from above. We rowed downriver and climbed that, too, over rocks thrown up by a volcanic event some 30 million years before.

Reembarking, we sped on a rollicking current to meet our first serious challenge: Eight Foot Rapid, so named because the river drops that far in a giddy space of about 125 yards. (The average fall of the river in this section is about 9.4 feet per *mile*.) We reconnoitered, with Millit pointing out each toothy boulder. "Stay away from the holes," said she in a firm schoolmarm's voice. A "hole" in the river is just that—water

rushing over a rock creates a boiling vortex into which a boat and its occupant can vanish. Wearing life jackets and surrounded by friends, we ran little risk of more than a severe dunking and a few bruises. I said something about swimming out of it. "Swimming has nothing to do with it," declared Millit. "The river will swallow you or spit you out." She grinned reassuringly. "Usually it spits you out."

We rowed into the maelstrom. Sand waves became jeroboams of water flung into our little craft. I was up to my waist in water, rowing mightily, when another boat, caught in a faster part of the current, collided with mine. My downstream oar caught on the rocky bottom with a banshee shriek, and I spun helplessly. Nancy, in the other boat, calmly rowed out of the disaster, leaving me free to pivot my craft and do the same. Bailing took a bit longer than usual—but it didn't matter, for we had arrived at our second campsite, and soon the spicy heat of a Mexican dinner took the ache from our bones. In one day we had climbed two mountains and shot three rapids. We told Millit, top sergeant by day and delightful companion by light of moon and campfire, that we'd remember this as boot-camp day on the San Juan.

Soon after sunrise, I found George Billingsley gazing off to the northwest at a dark smudge above the Abajo Mountains, perhaps 50 miles away. "It's snowing in the Abajos," said George. "Funny weather, high pressure coming from the north. Be about 80 today, down to 42 or so tonight. No rain." By now I had learned to believe in George's weather eye—he'd brought us through two consecutive dry nights.

Downriver a few miles, we paused to climb a series of tilted beds of limestone. Standing beside a rock no larger than the top of a bridge table, George placed a blunt forefinger on one fossil after another,

Winding westward toward Lake Powell, the San Juan River passes a series of sandstone spires and hogbacks formed by upthrust, tilted strata. The author's party climbed the Mule Ear diatreme—a volcanic jumble at right in the background—as well as the rocks near the camera.

127

Having conquered Government Rapids, most challenging section of the trip, author Charles McCarry rows triumphantly downstream. A hearty breakfast (opposite) fortifies the river-runners at daybreak. Below, the party enjoys a leisurely evening meal at its fourth campsite. The San Juan's water made delicious coffee, writes McCarry: "It was the proper color even before the coffee was added, and the grains of river sand mingled with coffee grounds to give the brew just the right texture."

crustacean and mollusk, that had lived in ancient seas that many times covered this land, receded, and returned only to recede again. Each time, the weight of the water compacted and cemented the bottom muck, entombing the creatures of the primordial soup. George invited us to count the fossils in a square foot of the rock. It was an impossible task. "I wonder," said George, who has the geologist's fatalistic view of change, "if any life form 250 million years from now will find trace enough of mankind to see and touch and wonder about us."

Some 20 miles downstream from our launching point, we had an exhilarating ride through six-foot sand waves near the Raplee anticline, an area of arching strata. Early expeditions, unblessed with our buoyant plastic boats, nearly swamped in such stretches. No one knows how the Indians who carved footholds up the sheer sides of the canyon walls fared in whatever craft they used. But the owlets we saw in a nest in a cleft in the rock, guarded by a mother with fierce yellow eyes, would need neither ladders nor boats in the free lifetime that lay ahead of them. That they would eat well we knew: Mice had visited us in our sleeping bags, and at twilight, bats soared through the mixed light of setting sun and rising moon, feeding on the abundant insects.

Until now, we had been utterly alone on the river. When Mexican Hat appeared, a hamlet at the edge of Monument Valley, we went ashore to replenish our drinking-water supply. I'd left all money behind in Bluff, but Jenny Urquhart loaned me a handful of coins tarnished by the river, and in the trading post I bought a bottle of pop. It tasted wrong after days of drinking little besides water, but the voices all round us speaking Navajo sounded exactly right.

BACK ON THE WATER, we passed through a great bend in the river and entered the most wonderful part of the San Juan—the Goosenecks, a series of loops in which the stream turns back on itself through canyons a thousand feet deep. The river seems to run more swiftly as it curves through the narrow Mendenhall Loop, circles the Tabernacle, and threads the Second Narrows, and we had little chance to look upward until we reached camp at dusk. As George made his ritual survey of sky and tested the wind, I saw the flash of sun on glass high above us on the lip of the canyon. A party of tourists was looking down from a roadside observation point. Their binoculars seemed as strange to me, and as much of an intrusion, as the telescopes of the first white surveyors must have seemed to the Indians who had lived for so long with shy spirits in these rocks.

That white men had a spirit of their own was vibrantly clear next day as I stood with my companions some 600 feet up the face of a cliff. Far below, the river's voice was a mere whisper, and my foot dislodged a rock that fell, slow as a gliding hawk, into the canyon. At last I saw it strike the ledge, but heard no sound.

We had climbed up a trail built near the turn of the century by gold prospectors and later named for one of them. The Honaker Trail is $2^1/_2$ miles long and ascends a cliff face 1,235 feet high. Its purpose was to link the canyon rim with a section of the river bank where a placer-mining operation was in progress. But the only packhorse ever to attempt the descent fell off the trail and left its bones to bleach on the rocks below. The narrow trail switching back and forth across the stone formation silently testifies to a remarkable engineering feat and the miners' strenuous effort—all for little benefit. *(Continued on page 137)*

Erosion-shaped contours of layered sandy limestone channel a tributary of the Colorado toward Marble Canyon, the long, slender northern entry to the Grand Canyon. Here a combination of little rain and thin soil permits only a sparse growth of cactus, agave, and other small desert plants.

PAGES 132-133:
Inspired by "these great mountains . . . the natural temples of God," the Mormon leader Isaac Behunin gave the name Little Zion to the area that became Zion National Park. Colorful walls near the Great White Throne rise more than 2,000 feet above the Virgin River, which carved Zion Canyon from a great formation of Navajo sandstone left by an ancient desert.

Morning sunburst breaches the crest of the Queens Garden in Bryce Canyon National Park. Queen Victoria waits at far left. Interbedding of siltstone and sturdier limestone has resulted in uneven erosion and fanciful formations. The trees, bristlecone pines, belong to the world's longest-living species: Its patriarchs surpass even the oldest giant sequoias.

Millit and Marilyn and Randy had been talking all voyage long about Government Rapids—the great challenge of the San Juan. At last we approached it, debarked, and looked down upon the rocks at its entrance, the rocks to the left, the rocks to the right. As we watched, a large raft and a canoe attempted the run, shipped several hundred gallons of water, and pulled in panic for shore. "On our honeymoon," said George, "my wife started through there in a kayak, turned upside down, pulled herself out, and bobbed all the way through, scraping the rocks as she went. Except for a few bruises, she came through all right."

So did we, by dint of keeping our eyes fastened on Millit in the lead boat and doing as she did. Once through the rapids, we were rewarded with the sight of a sparkling waterfall bounding down slopes of rosy Halgaito shale. Not far away, we noticed another sort of "waterfall"— a streak of black on the rock face. "Oil seep," George explained— petroleum oozing from solid rock. Early explorers found a pottery jug beneath this seep, giving rise to speculation that early peoples may have collected crude oil for fuel or medicine.

Marilyn let me daydream a bit as we passed into milder waters. One last rapid remained, and I did the unthinkable—I floated right into one of the dreaded "holes." My boat skimmed over a great rock, turned on its nose, and plunged into the vortex. The whirling water seemed as smooth as the skin of a trout. As it cascaded into the boat, stinging my face with sand and coating my glasses, nothing at all flashed before my eyes. Then the prow swung upward, and as I caught a glimpse of azure sky and pushed furiously on my oars, I knew I would keep my boat in the water and myself in the boat. As for the other two rules—why, they were worth breaking, if you could come that close to oneness with the river.

Blinded, I rowed ashore by ear, following Marilyn's voice. There I wiped my glasses, daubed the sand from my eyes, and shared a drink from my canteen with my young mentor.

We camped near the mouth of Grand Gulch, and when Nancy asked me what day it was, I was unable to answer—the sign of an ideal state of mind. Suddenly a wind rose. "Rain," said George. We donned waterproofs and pitched flapping tents. "Won't be much," said George, as chain lightning wrote sulfurous omens on a black sky far to the north. And, of course, there wasn't much—a brief spatter as if the wind were a boy spitting over the rim of the canyon. Then our companions Mars and Jupiter and Regulus appeared overhead to outjewel showy Venus off to the west.

There was little wilderness left as we entered our last day on the river, but it was wilderness enough. We found another waterfall, in a creek running deep and clean with rain that had fallen to earth far away, and we had a swim in a crystal pool.

Now Lake Powell came up the river to meet us, changing the color of the water from brown to blue—from Indian eyes to European eyes. For the first time we were in slack water, with no current to give us wings, no rocks to remind us of our fragility, no insistent voice of running water to fill our ears with the long, long song that began before the rising of our species, and will echo down this canyon after men and women have played their part in the scheme of things and vanished.

For our time in the canyon, and for our time on this journeying planet and in this abiding land, I—and, I think, all of us—carved on the stone of our memory the prayer we had read on the red rocks of San Juan Hill: "We thank Thee. . . ."

Western rattlesnake seeks shade beneath a rocky ledge. Opposite, hundreds of whimsical figures populate the Bryce Amphitheater. An early-day geologist termed such Utah formations "a singular display of Nature's art mingled with nonsense." Paiute Indians, more somber, saw these shapes as evil ancestors turned to stone by an angry god.

FOLLOWING PAGES: Sculptured over the ages by wind and sand, water and ice, dozens of heroic arches, domes, and pillars stud the landscape of Arches National Park in eastern Utah. Beneath dark storm clouds the evening sun sidelights Delicate Arch, whose open portal reaches 45 feet high.

*Preserved by the dry air,
the skeleton of a juniper
tree stands guard over
Canyonlands National*

Park in Utah. Below, winter clouds bring welcome moisture and a double rainbow to the remote and silent Canyonlands. A wondrous, stony expanse broken by cliffs, spires, and mesas, the park neighbors the former hideout of notorious outlaws Butch Cassidy and the Sundance Kid.

"Paved with rainbows . . . fallen to earth and
turned to stone," wrote one observer of Arizona's
Painted Desert. The eroded buttes of Blue Mesa in
Petrified Forest National Park owe their blue
stripes to traces of unoxidized iron. Spectacular
142   scenery typifies the Southwest's plateau country.

"It is, like any great beauty, the sum of small delights...."

THE PLATEAU

WHEN THE SNOWS MELTED in Wyoming in the spring of 1918, a handsome young cowpoke named Jack Fuss decided to drift down into what many people still called the Arizona Territory, though this vast region, filled with wonders and dangers, had become the 48th state six years before.

In Wyoming, Fuss had spent some time driving a horse-drawn bobsled between Story and Sheridan. Among his passengers was William F. Cody, who had shown Jack a necklace and asked if he could guess what it was made of. It looked like macaroni on a string to Jack, but Buffalo Bill explained that the trinket, which had once belonged to a Sioux warrior, had been made from the trigger fingers of 16 members of the 7th Cavalry who had died with Custer at the Little Bighorn.

At 26, Jack Fuss was well equipped for a life of adventure in the afterglow of the old West. He was a good shot, he could play the piano, he liked red men as much as white ones, and since the death of his beloved mother in Philadelphia, one place was as lonely for him as any other.

At a rodeo in Flagstaff, Jack made friends with a Cherokee Indian named George Blair. They decided to go up to Utah and spend a winter trapping. Jack bought three broncos from the rodeo stock, and after a certain amount of bucking and wahooing on these half-wild mounts, they headed north.

With the sharp white San Francisco Peaks at their backs, they rode steadily through regions of cactus and sagebrush, juniper and piñon pine. They saw little life with blood in it: a cottontail or two, a green lizard motionless on a twig. The country, heaved and wrinkled like an old brown hide, forced them northwestward.

Jack knew that the Grand Canyon lay between Flagstaff and Utah, but he didn't know exactly where it was. Nor had he imagined *what* it was. "We rode through a stand of cedars," he told me 62 years later, his eyes dancing with the memory. "And there it was. Oh, my gosh! Several thousand feet deep, I don't know how many miles across; we could see the river in spots. It was the biggest, most gorgeous thing I'd ever seen. It took my breath away, and I said a prayer."

George Blair said nothing. That night he vanished, a tall Cherokee wearing a big Stetson with a red hatband. Jack worried all night that his friend might have come to some harm—perhaps had got dizzy and fallen into the abyss. But next morning he followed George's tracks toward Grand Canyon Village and learned that he had taken a train to Williams. He never saw him again.

Maj. John Wesley Powell had explored this wonder of the world in 1869. After only a short time in the canyon, Powell wrote in his journal, "We are in our granite prison still." Jack Fuss remained for a year and a half, working with a Havasupai Indian called Chickapanagie in an asbestos mine and living in a cave. Was he ever lonely, ever bored?

Jack grinned. "The Colorado River didn't have a dam on it then," said he, "and it was as wild as a roaring tiger. It was so full of silt that if you fell in, your clothes would fill up and sink you.

"One clap of thunder would echo in the canyon for what seemed a half an hour, and lightning would run along the rim like a rattlesnake afire. I had work to do and plenty to look at, and old Chik was about as nice a fellow as ever lived. With him, everything was 'haniga'—that means 'okay' in Havasupai. I wasn't in prison. I was in paradise."

It is paradise still, or so I thought as I stood on Cape Solitude, 6,100 feet above the sea and 3,300 feet above the confluence of the Colorado and Little Colorado Rivers. Evening was coming down, and the Little Colorado, laden with minerals from the springs that feed it, was as blue-green as turquoise. The Colorado, according to how the light struck its surface, was the color of emerald or of topaz. These were the jewels of the geologist's clock that is the Grand Canyon. The hour hand, Precambrian rock at the canyon's bottom, stood at nearly two billion years; the

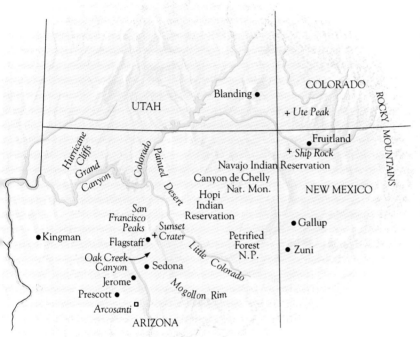

Reaching into four states, the plateau country sweeps north from the Mogollon Rim in Arizona to central Utah, and eastward from the Hurricane Cliffs to the southern Rockies in New Mexico and Colorado. Pungent pine forests yield to fantasy landscapes carved by the Colorado River and its tributaries. Past and future meet here as Hopi and Navajo cherish age-old traditions and newcomers explore vast mineral and energy resources.

minute hand, Cambrian rock, was at about 600 million years; and the sweep-second hand, the Permian rock that crumbled beneath the soles of my boots, was just passing 230 million years.

A sharp wind was blowing, and a few minutes before, on the canyon rim opposite Temple Butte, George Billingsley and I had amused ourselves by throwing flat rocks the size of silver dollars over the edge in order to watch the updraft, whistling up a cliff several thousand feet high, fling them back over our heads. George was reared on the Colorado Plateau, that great mass of eerily beautiful uplifted landscape drained by the mighty river at our feet. To him, rocks are not merely interesting, they are romantic. "I'll bet these rocks, lying here for millions of years," said George, "never dreamed that one day they would capture the Colorado, and it would make this canyon."

The Grand Canyon is certainly the most dramatic feature of the plateau, and most, like Jack Fuss, want to take a breath and say a prayer on first seeing it. But like any great beauty, the plateau is the sum of small delights; and as George and I turned away from the rim we saw blooming at our feet a bed of four-o'clocks, purplish flowers shaped like tiny bells that open in shadow in the afternoon and close when touched by the first strong light of the morning sun.

Crossing the plateau that day on little-used trails in blazing sun and blanketing dust, we had seen or sensed many another beauty mark on the face of the land: the magenta bloom of prickly pear cactus; the heady aroma of sagebrush; clumps of feathery (Continued on page 153) 145

Resplendent in turquoise and beadwork, two Pueblo Indians await their entrance at the Inter-Tribal Indian Ceremonial at Gallup, New Mexico. A jeweler from Zuni, Chester Mahooty (above) wears massive pieces made from stones he found years ago. He also fashioned the brass spear and the tomahawk, both used in tribal ritual. The feathered bonnet and beaded belt of Adam Trujillo attest to Plains Indian influence at Taos, northernmost of the pueblos. Taos Governor Trujillo makes drums and flutes.

Dismounting the hard way: A contestant in the saddle-bronc event at Gallup's Indian rodeo ends his ride abruptly—but fortunately without serious injury. Since 1922 the Ceremonial has brought together tribesmen from all over North America to perform and to display and sell their handicrafts. A parade through Gallup begins the day—a bit early for one young Navajo equestrienne. A small boy in elaborate Plains Indian garb marches behind her. Later, children of all ages tussle for treats in the "fruit scramble." A major tribal trading

center, Gallup styles itself the "Indian Capital of the World"—no mean boast in the Southwest, home of more than half of the Indians in the United States.

Morning mist drifts past mammoth coal-stripping machines at McKinley Mine at the edge of the Navajo Indian Reservation. The mine produces five million tons a year for utility plants and industrial customers from Texas to California. Opposite, steam-and-ash plumes rise from the Four Corners Power Plant at Fruitland, New Mexico. Though corporations have undertaken to reclaim stripped land and alleviate air pollution, environmentalists fear irreparable damage to the terrain and deplore the fouling of once pristine air. Perhaps a third of all strippable western coal lies on Indian lands. Jobs, tribal revenue, and new economic power come with development of this resource; but traditional leaders foresee harm to a cultural heritage that centers on the sacredness of the land. "O, our Mother, the Earth," begins an Indian invocation, "O, our Father, the Sky."

white Apache plume; swifts swooping through a deep, winding gorge; a Navajo boy alone in the fastness with a flock of sheep; two horsemen frozen on a horizon; the weathered logs of an abandoned hogan.

We saw no sparkle of water except in the rivers; yet beneath us, deep in the earth, were vast caverns and underground lakes filled with snowmelt and the rainfall that is gulped by the thirsty land in thunderstorm season. Entering one small fissure not far from where we saw the horsemen, we had basked in a cool wind blowing out of one of those caverns. Outside, the temperature was 90°F; inside the tiny cave, 57°.

From a vantage point atop Gray Mountain we had noticed, on the plain gashed open by the Little Colorado, what seemed to my casual eye uninteresting knobs. In fact, as George explained, these were the tops of subterranean rubble piles called breccia pipes, caused by the collapse of caverns some 2,000 feet below. "The interior mineralizes because water percolates down the pipe," said George. "Uranium and copper are commonly found in such features."

Riding proudly above the high-desert waves, Ship Rock points its sturdy prow toward snowcapped Ute Peak in Colorado. The 1,700-foot-high landmark near Four Corners—the only place where four states meet—and the lower geological dikes nearby stand as remnants of ancient volcanoes worn away by water and wind. The Navajo call the pinnacle "Rock with Wings."

Though the greater part of Arizona's mineral wealth has been mined in the desert—the silver lode at Tombstone, vast deposits of copper at Ajo and Bisbee, Globe and Morenci among others—the plateau has yielded its share, too. In 1898 the United Verde mine at Jerome produced more than 41.5 million pounds of copper; and it remained in production until 1952. Arizona is the nation's largest producer of copper—nearly two billion pounds in 1978, or 65.7 percent of the U. S. total.

A slight haze hung on the horizon as George and I gazed out on the plateau—the faraway smoke of a forest fire burning in the Coconino National Forest between us and Flagstaff. My mind sped back to Cima, California, high in the barren Mojave Desert. There, pausing for a cool drink at a general store that is Cima's only public structure, I had asked the storekeeper and postmaster, Mrs. Irene Ausmus, how she liked living there. "Oh, it's fine—if we don't get too many brakemen off the Union Pacific coming in and wanting soda pop all at once, like yesterday," she answered. "But what I miss is northern Arizona. We have *trees* there—*forests!* I was raised in the Hualapai Mountains near Kingman, but when I retire I'd like to live near Williams or Flagstaff, in the ponderosa pines and Douglas firs. I love the way they smell."

Unless someone like Irene Ausmus or George Billingsley helps a stranger out, he's likely to miss a lot on the plateau. I'd sought out the forests because of what Irene had said, and knew that Arizona's timber industry is an important part of its economy—and that its 11.2 million acres of national forest lands are a vital aspect of its environment.

Leaving Gray Mountain, George and I had encountered several curiosities. Isolated piles of stones, George said, had probably been left as Hopi religious markers. But scarecrow-like figures on juniper trees, in a region where I detected neither crows nor crops, had puzzled us both. Now came another mystery: From our campsite only a few steps from the edge of Cape Solitude, we observed globes of light gamboling in the night sky far to the east above a Hopi village where ceremonial dances were taking place. UFOs? Lightsome spirits released by the fervor of a deeply religious people? The spotlights of pickup trucks? It was good, chilly fun to speculate by our dying campfire, the only feeble point of light on the floor of this vast cavern of sky and stars.

The Hopi, numbering only 8,000, dwell in mesa-top pueblos or on the lower lands nearby, with the Painted Desert to the south and west, the Grand Canyon beyond—and Navajos all around. One of the Hopi

Budding apprentice at 18 months, Rajena Clouse watches intently as her father sets dowels into a rosewood drum top in a workshop near Prescott, Arizona. A native of Michigan, Danyel Clouse moved to the Southwest in 1975. He and his family produce musical instruments from imported hardwoods. Typical of their eclectic approach, they based one drum (opposite, below) on an Aztec instrument and embellished it with a motif from Maori mythology.

villages, Oraibi, is thought to be the oldest continuously inhabited community in the United States. There have been—as a young Hopi veteran of Vietnam named Valjean Joshevama, Jr., told me—good reasons for his people to withdraw to their mesas. "Once we occupied a wide area, but we were attacked by the Spanish, by the Navajo, by the Ute, and by the Apache," said Valjean. "We were few and they were many, so we gathered on the mesas and placed such warriors as we had—the Sun Clan, the Eagle Clan, the Fire Clan—on the perimeters. Still, the other tribes, and particularly the Navajo, pushed into our lands."

As early as October 1850 a delegation of Hopi elders journeyed to Santa Fe to complain to the new American authorities that the Navajo were raiding their orchards, cornfields, and villages for food. During the Civil War, when the Union recognized that its undermanned western garrisons could not deter Indian attacks on remote white settlements, Washington ordered Col. Kit Carson to round up the Navajo and relocate them. With the help of rival tribes, Carson captured many of the *Dineh*—"The People," as they called themselves—and marched them more than 300 miles to Fort Sumner in eastern New Mexico, thus writing the blackest phrase in the book of Navajo history: "the Long Walk."

In 1868 the Navajo who had survived the military campaigns against them and four years of captivity at Fort Sumner were settled on a 3.3-million-acre reservation in extreme northeastern Arizona and northwestern New Mexico. Soon the returning Dineh spilled out of this area, driving their flocks and herds before them, and occupied lands sacred to other tribes, including the Hopi. Bit by bit, by Presidential order and by act of Congress, the Navajo reservation was expanded until it became the largest in the nation. Today there are about 160,000 Navajo occupying 16 million acres, or 25,000 square miles.

Meanwhile, in 1882, as the Navajo continued to move westward, President Chester A. Arthur created a 2.5-million-acre reservation "for the use and occupancy of the Moqui [Hopi], and such other Indians as the Secretary of the Interior may see fit to settle thereon." At the time, about 300 Navajo were living—as trespassers, in Hopi eyes—on the Hopi lands. But despite Hopi complaints to government officials, the Navajo numbers increased, and by the middle of this century there were some 12,000 of them scattered over about 1.8 million acres of the Hopi reservation.

In 1958 the Hopi sued the Navajo. Four years later, a federal court ruled that although the Department of the Interior had not officially settled Navajo families on Hopi land, it had permitted such settlement to happen, and therefore the approximately 1.8 million acres of reservation on which Navajos had built their hogans and grazed their flocks were now jointly owned by the two tribes. In 1977, after repeated attempts at negotiation and mediation had failed, the land was divided in another federal court judgment. A partition line was drawn, awarding 911,041 acres to the Hopi and 911,041 to the Navajo. The 3,500 Navajo living on what overnight had become Hopi land, and the 50 Hopi dwelling on what had, between sunset and sunrise, become Navajo land would have until April 1986 to relocate. A family of four would be given $57,000 to buy a new home, along with their moving expenses; a bonus of $5,000 would be added if they moved without delay. The total cost to the U. S. Treasury is estimated at 250 million dollars.

To the outsider, this may seem to be a legal question, neatly if expensively settled. To the Hopi, who have been living on this land for at least a thousand years, and to the Navajo, who came perhaps three centuries ago, it is a religious question, and one that epics of conflict and epochs of prayer have failed to settle. "In the beginning, when all the Peoples emerged from the underground," Valjean Joshevama explained, "they scattered over the earth, and they were told to stop at whatever place they had reached when the Big Star appeared. Here is the place of the Hopi. Here is where they saw the Big Star. It is our eagle-gathering place, our wood-gathering place. It is *our* place."

But outside a Navajo hogan at Big Mountain, I listened to the murmuring voice of a woman named Mae Tso. Its pitch was so near to that of the evening breeze that I had trouble separating her words from the sigh of the wind. "My great-great-grandmother hid behind that ridge at the time of the Long Walk," said Mae Tso. "The soldiers found her and shot her, but she did not die. And when they let her go from Fort Sumner, she came straight back to this place, to no other. My nine children were born in this hogan. They were blessed with songs and prayers that are made with the names of these hills, the names of our animals, of our cornfields, of our spring. Now they say this is not our land. If they take us away, they will break our songs and prayers. I cannot go. If they take me from here, though I am only 43, I will die soon."

Dan R. Yazzie, a Navajo medicine man, sat on the dirt floor of the plyboard shack that has replaced his hogan: Since partition, neither Navajo nor Hopi can build new structures on land belonging to the other without permission. Permission is seldom asked or granted. Dan Yazzie told of increases in suicides, depression, and illnesses of spirit among his patients—a trend confirmed by doctors at nearby hospitals. While he spoke, his wife, Waldeen, sat on the floor of beaten earth and spun carded wool on a spindle. A suckling kid frisked between them, wagging its tail like a puppy. Grandchildren came and went, and the English words they uttered skipped across the rolling stream of Navajo like quartz pebbles flung into a brown river. Gazing to the east through the door of the hogan, Waldeen Yazzie said, "It is for the children that I am afraid. Away from here, there will be no harmony of life, no way of beauty for them."

Some families have already moved to Flagstaff and other urban areas, but traditional Navajos have found there the heartsickness Mae Tso and Dan Yazzie fear. Why, then, cannot the displaced Navajos find land on the vast tribal reservation that surrounds the Hopi lands? Because, the Navajo reply, precious grazing rights are handed down from parent to child; and because the land has already been badly overgrazed.

"Navajos talk of human rights," said Valjean Joshevama. "But they will not take their own people onto Navajo lands. They say there is livestock there already. They put livestock ahead of their own people."

As I LEFT HOPILAND and drove for hours across the Navajo reservation, with Venus over my shoulder and a waning moon throwing shadows on a landscape that is lovely even in the dark, I remembered something a friend of the Navajo once said to me: "The worst thing one Navajo can say about another is, 'He acts as if he has no family.' "

On the plateau, among all of its peoples, there is a strong sense of family. Dr. Florence B. Yount, an Iowan who came to Arizona with her classmate husband from medical school in Washington, D. C., to enter a practice founded by his father, speaks of the satisfactions of forging the chain of the generations.

Weather-chiseled arch called "The Window" frames rich bottomland and sheltering cliffs that for a thousand years have drawn inhabitants to Arizona's Canyon de Chelly. The prehistoric Anasazi, the Hopi, the Navajo all have dwelled here. But the rocky refuges could become death traps. The pictograph above recalls the 1805 Narbona Expedition during which the Spanish killed a hundred Navajo cornered on a ledge.

PAGES 158-159: One of the oldest ruins in Canyon de Chelly National Monument, Mummy Cave took its name from bodies found there. The Anasazi occupied the site from about A.D. 300 to 1300.

"I've delivered two thousand babies in Prescott and hereabouts," Dr. Yount told me. "I guess I got the most satisfaction from three 'pree-mies'—girls that came into this world as tiny little things we had to keep in incubators. Then you turn around and deliver six- or seven-pounders to *them* after they've grown up and married, and it makes you feel that maybe you've done a good job."

Prescott, lying in a pretty basin at the edge of a fragrant pine forest, was the first territorial capital of Arizona. There were a good many Confederate sympathizers in the Territory of Arizona, which had been created in 1863 when Congress cut New Mexico in two, and Abraham Lincoln wanted to be certain that the new government would be in loyal Union hands. When Mr. Lincoln's new governor, John Noble Goodwin of Maine, arrived in the territory in a howling snowstorm on December 27, 1863, he pressed on to the temporary site of Fort Whipple at Del Rio Springs, and a few months later founded his capital nearby on the banks of Granite Creek. The legislature named the new town in honor of William Hickling Prescott, who wrote *The Conquest of Mexico* and other great historical works.

From the day of its founding to the afternoon I first visited it, Prescott has considered itself a New England town planted in the wilderness, despite the Spanish and Mexican street names inspired by *The Conquest of Mexico*. Munching a noonday sandwich on the leafy courthouse square—the Plaza—I watched three old gentlemen play several games of checkers, and between the apostrophes of their cackles of victory, all indeed was Down East silence. On some of the older streets, stately 19th-century houses have been restored to their original grace.

"A printing press and a library arrived with the governor's party," Dr. Yount told me. "From the start there were lawyers, teachers, ministers of the gospel. Bands played and choirs sang. We had the first graded school in Arizona, in 1876. When John Charles Frémont came here as territorial governor in 1878, Mrs. Frémont, accustomed to the courts of Europe, found the ladies of Prescott the equal in charm and wit of any she had met in London or Paris."

Dr. Yount paused for a moment, and in the atmosphere of pluck and culture she had created, I was reminded of the Prescott woman, a schoolteacher, whose cabin was attacked by Indians while her husband was in town attending a session of the legislature. She sent her spouse a message reporting the incident and asking that he send more ammunition, as she had expended most of her buckshot driving off the hostiles.

"Our women were valued and our men deserved them," said Dr. Yount with finality. "They pass through my memory with colors flying. They helped build a town that has little to be ashamed of." Her mouth clicked shut for a moment. "Except perhaps," she said, "Whiskey Row."

Prescott's Whiskey Row, burned out with the rest of the downtown in a disastrous fire in 1900, was rebuilt and still exists, a line of saloons and shops on Montezuma Street facing the west side of the Plaza. And it has played its part in Prescott's history.

"The Palace Bar was the cowboy hangout," an old cowboy named Gail Gardner told me. "They didn't drink quite as much in the Palace as they do in the movies; but if a rancher wanted a hand or two, why, he'd usually find a couple sitting around in the Palace."

Gail Gardner, born of pioneer stock in the house where he now lives, tried the real New England. After attending Phillips Exeter Academy, he moved a little deeper into New Hampshire to Dartmouth,

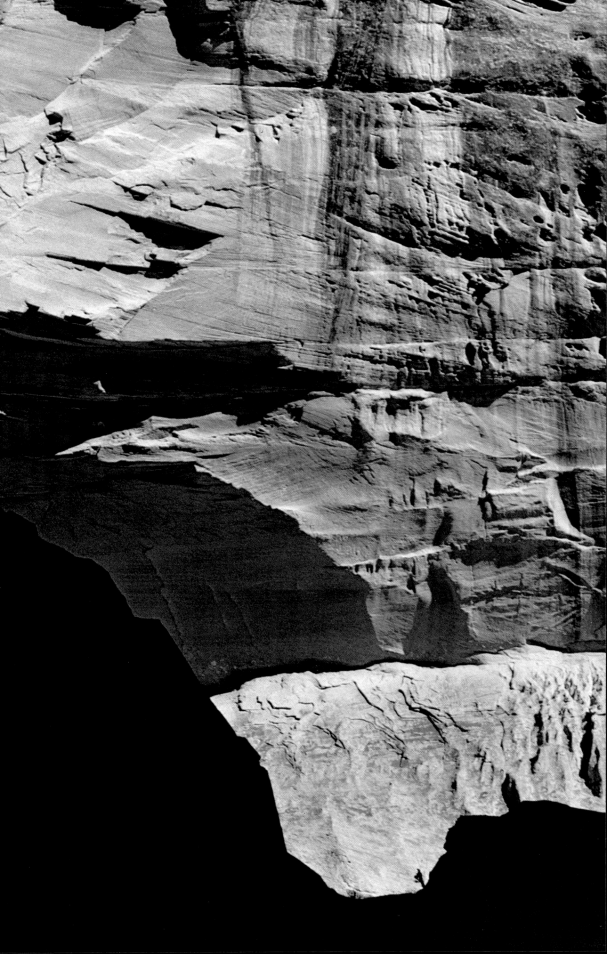

graduating in 1914. Then he got on a train for Arizona, and did what his heart told him to do: He began cowboying. With a partner, Van Dickson, he bought "a small cow outfit" in Skull Valley. He ranched there for nearly half a century, though a good part of that time he also served as Prescott's postmaster. "I'm 87 years old and I can't see or hear the way I could," said Gail cheerily, "but the forked end is still pointing down."

Talk about wild cattle! Gardner's ranch was mostly brush, and though his stock was Hereford, a lot of brindles and buckskins would come out of the woods with the whitefaces. Gail and the other horsemen, scrambling uphill or skidding downslope, would rope the mavericks with braided rawhide reatas, tie them to trees, and leave them overnight to cool off. In the morning, they'd lead them in one at a time.

"We'd use a running iron, or you could brand with a cinch ring held with two green sticks," Gail explained. "The iron has to be white-hot, so you fanned the fire with your hat—which is one reason a cowboy's hat looked so goldarn bad." In this rocky, wooded country, never penetrated by Spaniards or Mexicans who wanted to stay, the cowboy lingo nevertheless was Spanish—a roundup was a "rodeer," from the Spanish *rodeo*; a rope was a *reata*, a small herd was a *parada*, a long-ear or unbranded animal was an *orejana*. "Most of our cowboys came from California, and we were a clannish lot," said Gail, burrowing in a closet and showing me examples of items he was naming. "We used a single-cinch saddle, which was called a 'centerfire,' while down in Texas they used the double-cinch, or 'rimfire.'

"I rode into the morning sun for about 50 years; just liked working the cattle and knowing the country," said Gail. "My brand was the O N, and my earmark was a split in the left and a swallow-fork in the right. Early morning, you'd call to a bunch of cattle and they'd lift up their ears. You'd see which ones were yours. You're riding up in the country and you say, 'That old brockle-faced cow is about to calf—I better see if I can find her.' You had to savvy the cattle.

"The day of the range cowboy has ended. These rodeo hands are fine athletes, but most couldn't cut a cow out of the shade of the barn."

Gail himself, a lot of cattle folks told me, is headed for the Cowboy Hall of Fame in Oklahoma City. Years ago, he started writing what he calls "these cowboy poems"—verses about life on the range. Many a man recited a stanza or two to me, snorted, chuckled, and said: "Ol' Gail, he sure could write her down like she was!"

"My mother was real disappointed when I had this expensive education and went to punching cows," said the leathery old grasshopper they call the Cowboy Poet Laureate. "It helped when I was made postmaster. That was lots more respectable." He grinned in the joy of his memories—and I believe Mrs. Gardner, somewhere, was smiling, too.

Petrified Forest: Ancient rivers dumped thousands of conifer logs here, and they lay covered by alluvial silt and protected from decay for two hundred million years. Cell by cell, minerals replaced organic tissue, transforming each tree into colorful stone finally exposed by erosion. Lichens have taken hold on another log (opposite), beginning more changes as acid from the plants breaks down stone particles.

U P IN THE ARIZONA STRIP—a parched rectangle in the extreme northwest corner of the state, between the Grand Canyon and the Utah border—a homesteader named Bruce McDaniel recalled the look on his mother's face when she offered his older brother the first drink of water out of their well.

"Drilling a well was done by hand, with a tripod and a rope and a piece of sharpened pipe," said Bruce. "The deeper you got, the slower it got. Harold, my brother, had heard of a piece of tool steel over in the Hurricane Valley and he went off to get it. He'd been working a long time. While he was gone, my mother and I got to pecking away and got a little bit of water. She put it in a pitcher, and when Harold came back she said, 'Do you want a little drink of our well water?' " Bruce's guffaw blew us like a rain-filled wind back to that day in 1926.

Not that there is much of that sort of wind in the country the McDaniel family homesteaded near Moccasin. The average annual precipitation in the strip is less than ten inches. A section of land, one square mile, will support five head of cattle on the average.

"The seasons have changed for the worse," said Bruce. "Used to be you could just about figure it was going to burn up in June. But then in July, thunderstorms would come, and the grama grass would be 18 inches high. Now the last few years we don't get any summer rains to speak of, and this June grass or 'cheat grass' as they call it has come in. It saps the ground of its spring moisture, so the later grama grass won't grow. And when the cheat grass matures, it's full of sharp seeds that give the cattle big painful lumps in their jaws."

When the first pioneers came into this area in the late 1850s, they found native grass belly-high on a saddle horse and abundant water for stock at Pipe Spring. Pure and cold, it flowed from the Sevier fault, at what is now Pipe Spring National Monument, at a rate of 40,000 gallons a day. They also found bands of hostile Navajo. It was 1864, the Long Walk was in progress, and as Kit Carson himself admitted, thousands of the Dineh eluded him. The Navajo, whose lands generally lay southeast of the Colorado, forded the river and raided the settlements. Usually they took stock, but there was loss of human life as well, and many settlers left the country. Riding northward through a canyon near Zion National Park with my friend Melvin Heaton, I heard tales of the Navajo troubles.

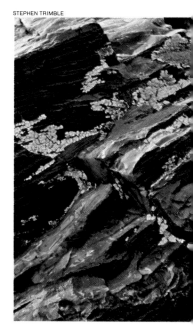

"The Navajo didn't like to take a life unless they had good reason," said Mel, who has spent time as a Mormon missionary to The People. "Right near here, a wagon train of Mormons was trying to move out of the country, and some Navajo shot their lead oxen. I guess they expected return fire. But all heads were cool as the Navajo kept on shooting arrows, except for one great big Scotsman. He got so mad the others had to tie him to a spare wagon wheel and throw him in the back of a wagon to keep him from getting them all killed."

In time the settlers came back, and their industry, combined with the population growth that accompanied the Mormon practice of polygamy (Mel Heaton's great-great-grandfather and great-grandfather, each with two wives, have 8,000 descendants between them), resulted in more intensive ranching and the overgrazing of what, in 1870, were called "the best herd grounds . . . that can be found in the Territory."

Bruce McDaniel may have described the Arizona Strip best: "Some fellow saw a jackrabbit, and he says to me, 'Get a gun!' I told him, 'If anything besides a rattlesnake, a mustang, and me can live in this chewed-down country, let it be!' "

Let it be. All over the plateau, and indeed all over the Southwest, I heard that cry, and the words: "You can't change one thing without changing something you don't want to change." There is widespread resentment toward what is often seen as bureaucratic meddling.

"The 13 western states," said Calvin Black of Blanding, Utah, "have about the same relationship with Washington as the 13 original colonies had with London.

"Washington sends out dream-filled bureaucrats from the Forest Service, the Park Service, the Bureau of Land Management, the

Department of Energy. They won't listen to the local people. But we have to live with their mistakes."

Calvin and I had just emerged from a disused uranium mine bored deep into Red Canyon, in San Juan County. Inside, the blackness had swallowed the beam of my flashlight, like the mouth of a shark taking a swimmer's pale leg. Now we were standing at an elevation of perhaps

Chapel of the Holy Cross soars high above its red-rock pedestal near Sedona, Arizona. The donor, Marguerite Brunswig Staude, devoted 25 years to the planning and building of the concrete-and-glass memorial to her parents. The Roman Catholic sanctuary welcomes people of all faiths for meditation and silent prayer.

8,000 feet, and we could see for a hundred miles. Calvin spoke of the uranium strike that in the 1950s brought a moment of prosperity to the Hole in the Rock country. With pick and shovel and wheelbarrow, Calvin mucked uranium ore out of ancient riverbeds now deep underground. He saved his earnings and invested wisely.

To the west we could see the rosy rim of the unspoiled wilderness of the Kaiparowits Plateau near Bryce Canyon National Park. Outsiders had defeated the construction of a coal-burning power plant there. Calvin Black, savoring the pure air we were breathing, nevertheless mourned the loss of economic opportunity. He spoke of ever more stringent regulation of mining, oil production, natural gas. "Eliminate those three things and you eliminate 85 percent of our tax base and 60 percent of our jobs," said Calvin. "We've lived poor in this country for a hundred years, and seen our increase leave because there was no work for them. Then at last we got enough money to build roads, and strangers came in and said, 'You can't have Utah—it's too beautiful for such as you!'"

Calvin Black is one of the leaders of a political movement called "the sagebrush rebellion." Simply put, the movement's aim is to turn over a large part of the vast federal land holdings in the West to state control. Indian and military reservations and national park and forest properties are not at issue at present; but Nevada, backed by Utah and several other states, has prepared a lawsuit against the federal government contesting control of millions of acres of other lands now administered by the Bureau of Land Management.

In Arizona, the legislature had to override Governor Bruce Babbitt's veto in order to support the suit. Like many others, the young governor believes such a move by the states to be unconstitutional and unwise. "Public land is a public bank account," he said to me. "It provides an enormous opportunity for us to develop it for the good of the people. As to the sagebrush rebellion, its leaders know in their hearts that it's just puffery."

Perhaps they do. Calvin Black told me: "I know I use strong language. I've learned that if you don't shriek, America won't lift its ears."

That is an image that might bring a smile of agreement to the craggy face of Peter MacDonald, chairman of the Navajo Tribal Council. "The goal of the Navajo Nation," he told me, "is to control its resources and generate jobs so as not to rely on the federal government to carry us from month to month." Chairman MacDonald spoke of a period during his boyhood when a federal program to control overgrazing drastically cut the size of Navajo livestock holdings: "My grandfather's large sheep herd was reduced to 85, his cattle to only 10 head. Our horses were slaughtered. It almost broke the back of the Navajo. I went away to school because I had no sheep to herd."

Peter MacDonald and many others like him came home from school determined to take the fate of their people into Navajo hands. The discovery of coal, uranium, oil, and other minerals and energy sources on the reservation has given them their opportunity, and if there is a sagebrush rebellion, there is also a hogan rebellion. "Until now, 90 percent of the energy derived from resources of the Navajo reservation—150 billion kilowatt hours—has gone to the outside world," said MacDonald. "Henceforth, we will take the first benefits and share the remainder on fair terms. That is the policy of the Navajo Nation, which was left with nothing after the Long Walk, and left with nothing again when they took away our livestock."

Obviously there are many questions about what lies ahead in the great Southwest. Upon entering the plateau, 70 miles north of Phoenix at a place called Arcosanti, I caught a glimpse of one man's vision of the future. On high grassland surrounded by amber rocks and blue hills, an architect-engineer-philosopher named Paolo Soleri is building a city. So far, it is only a table model—though several full-scale structures have been completed—of a geometric form of glass, concrete, and steel that will enclose a steep hillside.

Greenhouses will provide food and climate control. Each inhabitant will have his private space within. High-speed elevators will whisk him to theater, concert hall, school, physician. If he wishes to go outside into the natural climate, he may do so, and of course he will be able to see the grand surround of nature through the glass walls of Arcosanti.

Paolo Soleri believes in transformation of the environment, in technology, in frugality. "Something has to change," he says, and his disciples believe that the change will come in his city. "Lately I've been told I'm more a theologian than an artist," Dr. Soleri observes.

With luck, the city will be completed by the turn of the millennium. I asked Tony Brown, a young English architect who showed me around the site of Arcosanti in Soleri's absence, if the builders of this utopian place believed that an environment designed in a certain way would produce certain behavior in human beings. "Oh, yes," said Tony. "Paolo is very much into the plumbing of society; we must recognize the power of the city to influence our behavior."

As I drove away, I remembered the hill towns of Tuscany which seem to grow out of the very earth, and it occurred to me that Paolo Soleri, Italian-born, was perhaps building an Assisi under glass. I felt certain that he will take care to leave a window open so that the birds of St. Francis may fly into his city.

Spring moon rises over Oak Creek Canyon as the retiring sun leaves a golden glow. Spectacular formations and

easily accessible recreational areas draw visitors by the thousand to the 12-mile-long chasm.

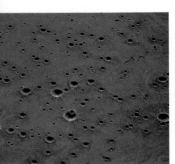

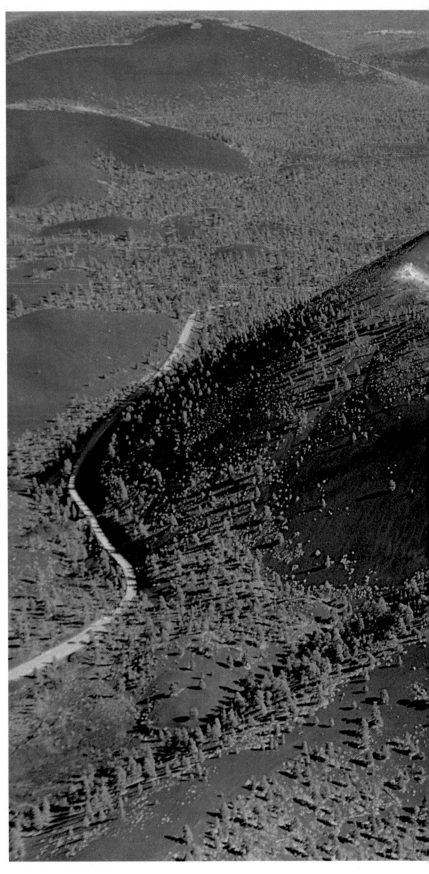

Fiery eruption 900
years ago thrust
Sunset Crater high
above the surrounding
plateau. Its distinctive
red tinge, from
oxidized iron, earned
the thousand-foot-high
cinder cone its name.
Ponderosa pine trees
vie with wind-swirled
cinders to blanket the
sloping sides. The
mound marks the most
recent activity in
a field of 400 volcanoes.
When NASA needed
a moonlike surface to
test Apollo vehicles
and their astronaut
drivers, the Geological
Survey blasted out
craters in a nearby
cinder field (above),
some measuring 45
feet across.

*Haloed by the setting sun, spiny, ribbed giants
stand sentry at Saguaro National Monument
west of Tucson. Familiar and distinctive symbol of
the Arizona desert, the saguaro may attain a
height of 40 to 50 feet and a weight of nearly ten
tons in two centuries of slow, irregular growth.*

"Among the saguaros were myriad quail; we could hear their murmuring...."

THE DESERT

O N THE SLOPES of Picacho Peak between Phoenix and Tucson, and as far beyond as the eye could see from an airplane 2,000 feet above the Sonoran Desert, the giant saguaros stood rank upon rank like figures of chivalry.

Among the troops of cactus ran a great fissure in the desert floor. The scar was nine miles long, and it marked one of the many places in Arizona where the soil, drained of its underground water by wells, is shrinking and cracking. The accepted and less dramatic term for this phenomenon is "subsidence." The shadow of our Cessna raced across the primary causes of the problem: perfect rectangles of dark soil planted with greening crops; a maze of mobile homes beside a river of concrete flowing with four glittering lanes of traffic; and, to the south, bejeweled by emerald swimming pools, brawny Tucson, striding steadily north into the desert while Phoenix marched south in the burning sun to meet it.

In most cases, land subsidence is a result of the energy, the confidence, and above all the thirst of modern man. Observed from the air, his works stir the heart with pride—that fan of irrigation water, coaxing flawless produce out of arid earth; the crop-dusting plane below our own aircraft, trailing a festive plume of insecticide spray; the hundreds of men and women in cars, heading home from good jobs to swim or relax in air-conditioned houses. After all, what a wondrous thing is man, to have raised this active, sparkling civilization on the floor of a dead sea, in the blue circle of mountains that were its shoreline and its islands!

"If you take a lifetime to mean 60 years," I was told in Phoenix by an eloquent government scientist named Richard Raymond, "then you can say that man has lived on this part of the earth for nearly 200 lifetimes. Thirty lifetimes ago, the Hohokam farmer learned to irrigate crops. One lifetime ago, his modern successor began to use the power-driven pump, which permits him to draw water out of the earth faster than rainfall can replace it. And in that brief span of time, man has lowered the water table by as much as 450 feet."

Raindrop by raindrop, water has collected over the eons in the basins between the mountains. The deeper it lies, the older it is: Some people in Tucson are drinking water that fell from the skies 10,000 years ago. Whereas Phoenix and its valley are partly supplied by the impounded surface waters of the Salt and Verde Rivers, the Tucson area is wholly dependent on groundwater sources. But water in underground reservoirs serves, among other things, a special purpose: It holds up the earth. As water is pumped out to grow crops, fill swimming pools, supply industry, and keep the hair of lovely Arizona girls shining in the sun, the earth's surface sinks. In one place studied by the U. S. Geological Survey, the ground subsided 12½ feet between 1952 and 1977.

Said Professor Sol Resnick of the University of Arizona, "We're using twice as much water as we get back from nature. But there are solutions. We're going to channel water to Tucson from the Colorado River. It's true that the water in the Colorado is oversubscribed—but remember that 90 percent of its flow comes from the western slopes of the Colorado Rockies. We can increase snowfall there maybe 10 to 15 percent by seeding clouds. Lake Mead and Lake Powell, the biggest reservoirs on the Colorado, together lose an estimated 1.2 million acre-feet a year by evaporation. If we reduced the surface temperature by 3°—say, by circulating the deeper, colder storage—we would save twice as much water as Tucson uses in a year. If the doomsayers would avail themselves of existing technology, they wouldn't have to worry about water."

Professor Resnick, like geologist Raymond of the Water and Power Resources Service of the Department of the Interior, is among a small company of scientists who are trying to understand the scope of the problem. They admit they have a long way to go before they can offer definitive answers. As Sol Resnick put it in his staccato fashion: "Nobody really knows what's down there."

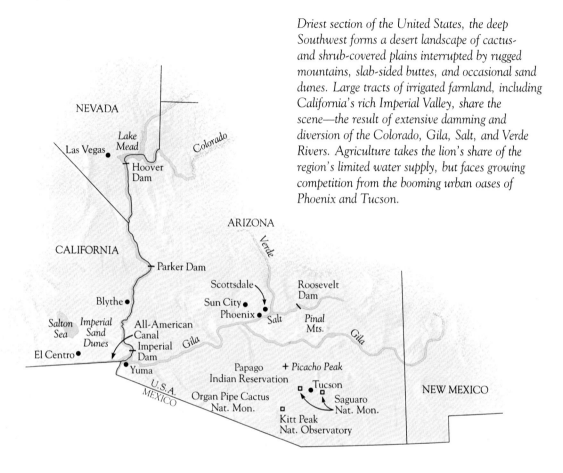

Driest section of the United States, the deep Southwest forms a desert landscape of cactus- and shrub-covered plains interrupted by rugged mountains, slab-sided buttes, and occasional sand dunes. Large tracts of irrigated farmland, including California's rich Imperial Valley, share the scene—the result of extensive damming and diversion of the Colorado, Gila, Salt, and Verde Rivers. Agriculture takes the lion's share of the region's limited water supply, but faces growing competition from the booming urban oases of Phoenix and Tucson.

A few days after my encounter with the scientists, I sat in silence among the saguaros on the Papago Indian Reservation. With me was a Papago named Daniel Lopez, and from first light Danny and I had been wandering through the majestic forest of cactus and arid-land shrubs that white men call the Sonoran Desert. Now it was last light, and something marvelous was happening on the western horizon. The sun, like a round face painted with ochre, was going down—and just above its forehead hung the new moon, a crescent of hammered silver.

The hours had been filled with the rippling, melodious sound of the Papago language. As I waited for our mesquite fire to burn down to cooking coals, I remembered some of the day's events, and one in particular: Walking, we had come into an abandoned village, our footsteps deadened by the grass of the desert floor. Wind sighed through the walls of empty huts made of slender stalks of ocotillo, a thorny plant that is the natural barbed wire of the southern deserts. Hidden among the thousands of saguaros, some more than 40 feet high, were myriad quail, and we could hear their murmuring. My heart leaped at what I saw next. In a patch of shade, three haughty coyotes took their ease with lolling

Searching the sky from its domed, 19-story-high housing, the Mayall Telescope focuses on the outer universe at southern Arizona's Kitt Peak National Observatory. Located on a mountain site leased from the Papago Indians, the observatory boasts the world's largest concentration of sophisticated astronomical equipment.

tongues. The breeze ruffled their gray coats, and even in their stillness one could sense the speed hidden in the muscles of their lean bodies.

The Papago, with their cousins the Pima, once ranged over 18,000 square miles of arboreal desert in what is now called Arizona and Sonora. Today some 15,000 of these gentle folk live on or near three reservations totaling 2.8 million acres—more than twice the area of Delaware, and the largest Indian reserve in the United States except for that of the Navajo.

The Papago were some of the earliest North American Indians to be christianized by Spanish friars, who settled among them in the late 17th century. The gleaming Mission of San Xavier del Bac, white as the bones of a saint, stands on the reservation at the southwestern outskirts of Tucson. Its first burnt adobe brick was laid about 1778.

The Papago harvested at least 200 desert plants. When the rains came in spring, they moved from villages in the mountains to the lowlands and impounded water behind simple dams to raise corn and squash and beans. In their domain there was no flowing water except for flash floods. (For centuries the annual rainfall in the Sonoran Desert has averaged less than 15 inches in the east and less than 5 inches in the west.) The Papago hunted the sparse and wary game: white-tailed deer, jackrabbit, sometimes quail. A warrior who killed an enemy in battle drew apart from others in his village and purified himself in a four-day ritual.

The number four figures in almost everything these people do—because, Danny Lopez explained, of the four directions. Their songs and dances praising nature or describing the ferocity of ancient enemies are performed in four verses or movements.

"Our people were great runners. We hunted with the bow," Ramon Lopez, Danny's father, told me. We were seated in the shade of the ramada of ocotillo outside his adobe house, enjoying a cool glass of water. Earlier that morning I had come upon Ramon Lopez as he hoed a cactus garden in his yard. In it were a saguaro, a cholla, a barrel cactus, a mesquite—an individual specimen representing each of the species that grew in such abundance for miles in all directions. Ramon Lopez had his own unspoken reasons for linking his life with the life of these plants.

Some mornings, in breechclout, Ramon runs several miles in the desert. He also practices with his bow. The feather at its tip flutters as he runs, and he sees its shadow flickering before him. He showed me a sphere of mesquite wood—the kickball used in a Papago game. "It is one man against one man. You do not kick," said Ramon, "you lift the ball on the toes and throw it ahead, one man racing the other to the ball to throw it ahead again—maybe 15 miles out and 15 miles back. Oh, we would bet on that game; people would bet their clothes and, if they lost, go home almost naked. When I was a child, a man from another village came and told us he would win. We all bet on him and he lost. We were amazed, because we could see he was not a good player. But we had never met a liar before.

"Running meant much to the old people," mused Ramon.

"He is running to join them," said his son in a quiet voice.

Danny is working to join the old people to the young—reviving the Papago language in the reservation schools, helping children to learn the songs and dances. The Papago believe that some people are gifted, that in their sleep a talent for music or storytelling is given to them. One such may be 8-year-old Juanita Havier. Her mother's mother began teaching her songs when she was an infant, and now she knows many. I heard her sing them, and one I remember well, "The Song of the Morning Star." As Juanita sang, with the growing things of the desert in silhouette against the sky and the sacred mountain Baboquivari on the horizon, she shook a gourd rattle, Danny drummed the same rhythm, and another accompanist brought sound out of a notched stick. On her head Juanita wore a crown made of the famous basketwork of the Papago. Woven into it was the morning star—the woman's star, for she must rise with it to begin her long day's labors. That was what Juanita's song was about: the carrying of water in an earthen jar in a burden basket, the grinding of corn with *mano* and *metate*, the harvesting of the fruit of the saguaro, the gathering of bear grass, yucca leaves, and black devil's-claw fibers to make the coveted baskets.

No longer do many of the Papago work the soil. Virtual disappearance of their traditional culture has brought them confusion, disorientation, and unemployment. Among the flowers on the desert floor, no matter how far we traveled on the reservation, Danny and I came time and again upon a more sinister harvest: great heaps of empty brown beer bottles. "I weep to see old people and young people drunk in our villages," said Danny.

"Not long ago, this was never seen among the Papago," added Ramon Lopez. "We raised food, and if we needed a little money for clothes for a child or for the medicine man, we sold a cow. Each person when he woke knew that he would do something useful that day—plant a seed, run a race, tend a crop. Now he waits. . . ." Ramon put away the bow he had been showing me; the gesture seemed to say that for the Papago, life had changed forever.

Even Danny Lopez's Papago name was inspired by the white man's world. He is called Wainami O'odham, Iron Man, because he saw a movie about robots as a child and, when he returned home, imitated the clanking creations of Hollywood. But Danny, like Ramon, still runs in the Papago way. Earlier this year, he competed in the marathon at Tucson. "At the end," he told me, "when my heart was bursting and my legs were going to break, I saw in my mind the faces of the children, and of the old ones, too. I went on—because they sang to make me strong."

A cousin of Danny's had chatted with me for a moment on the beaten dirt square of their village on the reservation. Felix José knew what was in the mind of Wainami O'odham. "Danny thinks a lot," said Felix, pointing with his chin to Baboquivari where I'itoi, father of the Papago, waits to help his people should they send a runner to him in time of trouble. "Danny thinks about that mountain," said Felix José.

ADAY LATER I stood in a window high in the Arizona capitol, gazing out on Phoenix, that vibrant combination of Babylon and spaceport that has grown in population from 106,000 in 1950 to nearly 800,000 in 1980. Beside me was another man who thinks about mountains—Bruce Babbitt, the energetic young Governor of Arizona.

"This is the oasis society," said Governor Babbitt. "It is futile to try to discourage growth. Free people have a right to go where they want. It used to be that workers went to places where there were factories. Now factories come to places where people want to be. We welcome our newcomers, and we're strengthened and refreshed by them. The question is: How do you accommodate them without wrecking the whole thing?"

Bruce Babbitt, descendant of Arizona pioneers, has much in common with Danny Lopez. Both feel the beauty (Continued on page 180)

Orange groves studded with frost-abating windmills (opposite) stretch before the Picacho Mountains' Newman Peak, northwest of Tucson. Once dependent on impounded river water for irrigation, Arizona farmers now draw the greater supply from underground. But as the water table drops, the cost of pumping soars. Overpumping may actually collapse large sections of land and, along with erosion, open up miles-long fissures (above).

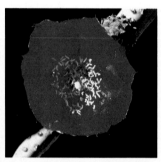

Wild flowers brighten the Sonoran Desert in early spring and again after midsummer rains. At right, poppies and lupines color slopes of the Ajo Mountains in March in Arizona's Organ Pipe Cactus National Monument. Desert annuals require above-average winter rainfall to bloom spectacularly; seeds of some species may wait out several dry years before germinating. Cactuses usually blossom in April and May, after the annuals have died and competition for pollinators has become less keen. On this page: rainbow cactus (top); Santa Rita prickly pear (above); beavertail prickly pear (above, right); and pincushion cactus. Cactus plants store water, so they may flower even during droughts.

Threatening thunderclouds gather above a classic multiarmed saguaro. Bearing seeds and fruit that nourish many animals, the stately cactus harbors woodpeckers and flycatchers in holes the birds drill into its fleshy body. Shallow, wide-spreading roots—also a feature of the prickly pear (foreground)—soak up surface water for storage in the plant's succulent tissues. At top, a collared lizard surveys his surroundings. Above, a carpet of owl's clover frames a hedgehog cactus. Enlargement of a hedgehog flower pistil, below, shows details of its green stigma, pink style, and yellow stamen.

of their native land like wind on the skin; both wish to preserve their people from the fate of the Hohokam, an advanced civilization that flourished in southern Arizona from about 300 B.C. to A.D. 1400. The Hohokam—in the Piman language, "Those who have vanished"—created an irrigation system on the Gila and Salt Rivers that is still, in a sense, in use today, for the canals built in the last century by the white settlers of Arizona followed the courses of those laid out more than a millennium ago. The Hohokam produced pottery, jewelry, and a political system that worked for almost two thousand years, then disappeared.

"My grandchildren must not live in a withering civilization," said Governor Babbitt firmly. "I'm not as pessimistic as many environmentalists. I am not going to leave office until we have addressed the groundwater question. I think we can achieve safe yield within a few years—that is, a balance where we take no more water out of the ground than nature returns to it. I have introduced legislation that will enforce mandatory conservation by everyone. I know that the people will support a reasonable program in which everyone shares the burden."

Within a few weeks of our conversation, the Arizona Legislature passed the Groundwater Management Act of 1980, a far-reaching code that puts strict controls both on residential and commercial land development and on agricultural wells. The new law does divide the burden, but as was inevitable, the major portion will be borne by Arizona's farmers. Agricultural irrigation accounts for about 90 percent of Arizona's water usage. Already the falling water table and increasing pumping costs have made it uneconomical to cultivate marginal lands.

Professor James Becker, an agricultural economist at Arizona State University, points out pressures other than water shortage behind the removal of land from agricultural use. As the growing population demands housing, more and more land is subdivided by developers. "A 3,800-acre ranch north of Phoenix was bought by speculators in July 1979 for $8,000 an acre," Dr. Becker said, writing the figures with a fingertip on the table between us. "In March 1980 it was sold for $11,289 an acre. That is a 40 percent profit in eight months.

"The basic philosophy in Arizona has been to do nothing to restrain anything if a dollar can be made. On what used to be prime agricultural land between here and Tucson, you now see signs: *Dust Storm Alert.*" Thousands of acres of farmland go out of production in Arizona yearly, while thousands of glassy houses, shining like shards of a broken mirror scattered on the sands, rise in their place.

"Imagine what that means to an Arizona farmer," said Jim Becker. "He has moved rocks, worked all the summers of his life in temperatures up to 110 or 115°, seen crops and topsoil wash away in floods, watched crops die in drought. But he's proud that he's survived, and finally he's ready to pass on his farm to his children. Then his son says, 'Dad, you can raise cotton if you want. I'm going to sell this land to a developer and raise mobile homes!' "

NOT ALL THE NEW HOUSES appear on former farmland, of course. Homesites range from the steep sides of red buttes to rocky canyons all but unaltered by man. Twenty miles northeast of Scottsdale I called on Fred and Donna Corbus and their two children, 9-year-old Kienan and Melea, 6. Better than any other object I saw in the desert, the Corbus home summarized for me the ideal relationship between the works of nature and the works of man. Designed by a young Phoenix architect named Edward Bosson Sawyer, it is a glass cube—one big room, without partitions, without even a wall on which to hang a painting. The house is set among huge granite boulders, and looks from the outside like some space-age machine that has landed on the surface of a planet in order to gouge samples from its virgin surface. But from the inside, it is a window upon nature—in its construction hardly a stone was disturbed or a plant uprooted. It is heated by solar energy, cooled only by the wind. The Corbuses sleep on the floor, and sit on the floor to dine. Birds and other animals come and go on the gray rocks just outside the windows, and storms form over the cones of the mountains to the north.

What does it give the Corbuses, who built much of this house with their own hands? "Magic," says Donna Corbus. "Romance."

For architect Ned Sawyer, who played as a child in the famed Boomer house by Frank Lloyd Wright while it was under construction, the Corbus house has won state, regional, and national awards. "The Corbuses are living in a sculpture," Ned corrected me when I called their home a machine. "The work of an architect should say : 'Man has placed this here. But his hand was skillful.' "

It was a deft hand indeed that placed the Arizona-Sonora Desert Museum among the saguaros west of Tucson. That hand belonged to William H. Carr, co-founder of the museum with Arthur Pack. On a 100-acre tract near the Tucson Mountain Section of Saguaro National Monument, William Carr created not a sculpture but a living mobile in which birds and mammals, reptiles and insects, cactuses and trees and shrubs move ceaselessly within the great gallery of the Sonoran Desert.

Early one morning, while a tuneful wind cooled the newborn day, I sat in the aviary and watched a Gila woodpecker carving out a nest within a giant saguaro. This cactus can live as long as 200 years and achieve a height of 50 feet; at maturity, saturated with water, it is likely to weigh nearly ten tons. Like most desert organisms, it is patient: At 3 years of age, it may be a mere 2 inches tall; at 15 years, 18 inches. Not until it is 50 to 100 years old do the first arms appear.

I heard a laugh of delight, the sort of chuckle inspired by the antics of a mischievous child, and then a voice with a light Viennese accent said, "That woodpecker is making a terrible hole in the saguaro, and he is getting sticky feathers. We shall have to give him a bath."

The speaker was Dr. Inge Polgayen, former student of the naturalist Konrad Lorenz and now curator of birds and mammals at the museum.

"The Gila woodpecker and many other birds could not live in our hot summers without the insulation of the saguaro," she said. "When he abandons his nest, it will become a home for flycatchers and owls. The jackrabbit rests in the shade of the saguaro—his enormous ears are a network of blood vessels, and the wind blowing on them cools his whole body. Oh, desert creatures are fascinating!

"Soon the white-winged doves will arrive after their winter in Mexico," she continued. "They come in clouds, and feed on the nectar of the saguaro flowers and pollinate them. Later they eat the fruit and disperse the seeds in their droppings on the desert floor." By night, nectar-feeding bats migrating from Mexico also visit the saguaro.

Dr. Polgayen came to the Southwest in 1955 with her husband, Ivo, who is director of the Reid Park Zoo in Tucson. Neither longs too much for the Vienna Woods. "We visit Austria and then we come back here where the people are open and the country is wild, and we say, 'We are home; here are the rocks and the desert again!' "

Transparent gem in a granite setting, the home of Fred and Donna Corbus near Scottsdale, Arizona, reflects their belief that "we are guests of the desert." Redwood decks and an open interior permit the family to share the company of the area's abundant wildlife. Offset from the house, solar collectors heat water and space; canvas drops, screens, and vertical louvers shade the breeze-cooled structure. "The feeling," says Fred Corbus, "is truly one of living in a sophisticated tent."

Heading west, I crossed the line into California and drove through the great dunes of the Colorado Desert. Across the Imperial Valley at Plaster City, my ears, lately accustomed to the silence of near-wilderness, were assaulted by the roar of some 300 motorcycles assembled from urban southern California for an enduro on the desert sands.

"The noise drives me crazy," screamed my companion, Kay Decker, of the Interior Department's Bureau of Land Management. "But I must say I like these folks—they leave the desert clean, and it's a family event, always."

As Kay spoke, a small boy on a small motorcycle zipped between us wearing a fiendish grin and leaving behind puffs of blue smoke as he ran through the gears and blew carbon off his spark plug. "I'm afraid they don't like us," Kay added. "They think BLM is closing the desert, and they resent it."

ORV enthusiasts—O-R-V is short for off-road vehicles, and the term can make some environmentalists' blood boil faster than it can make a motorcyclist's heart leap—have been blamed for widespread damage to the desert. Their vehicles leave tracks on the natural terrain that may not heal for centuries. They have been accused by their adversaries of vandalism, thuggery, and worse.

Bob Halsted, a San Diego insurance agent who is vice president of the Busheaters Motorcycle Club, bitterly resents that image. "This is my desert, too," he said, "and my family and I have a right to use it. I come out here and camp after five hard days in the city. I see that sunrise in the calm and cool, and I feel in love with this place."

BLM is developing a plan that might classify as wilderness or as limited-use area more than half of the 12½ million acres of public desert land it manages in California. That would restrict access to vast tracts of this bleak territory for people who have always been free to roam on it, virtually unfettered by regulations. Already certain areas have been closed to protect archaeological sites and rare species of plants and animals. One such is the flat-tailed horned lizard. "I don't worry about that lizard," said Bob Halsted. "That fellow has lived here for centuries and survived. I can't believe a motorcycle will hurt him."

ORVs participating in organized off-road events are now confined to a small area of the desert, though individual machines may travel any unclosed road. "If you let unregulated ORV activity spread," said Gerald E. Hillier, BLM district manager, "you will lose the very characteristics that everyone values: solitude, scenery, the uniqueness of the desert country. For years, many people viewed the desert as a place to dump old bedsprings and wrecked cars. Now the public knows it is valuable—for recreation, for mining, for agriculture, for its potential as a source of geothermal energy, for uses we haven't even imagined. But there are competing considerations—it would be equally as wrong to kill off a rare species as to prevent the development of an exciting new energy source."

There was excitement in the hearty voice of Norm Hodgdon, manager of a geothermal pilot plant being built at East Mesa in the Imperial Valley. "What we've done here is make electrical power from the fluid in relatively low-temperature thermal wells," said Norm. "Flash systems take steam out of the earth to turn a turbine, but put back less than 100 percent of the water they take out of the ground. We use a binary system—heating a second fluid, such as propane or butane—that permits 100 percent of the water to go back underground."

Hodgdon's plant, built by Magma Electric Co., is designed to produce 10 megawatts of power. "The Geothermal Research Program of the U. S. Geological Survey estimates that there are about 6,800 megawatts of identified electrical potential in the Imperial Valley geothermal systems," said Dr. Wendell A. Duffield of USGS. "By my rule of thumb (1 megawatt is sufficient for a town of 1,000), this could supply the electrical needs of a population of about 6,800,000."

The electricity, said Norm Hodgdon, can be sold for ¼ cent a kilowatt-hour—cheaper than electricity from most other types of generators. The geothermal plant pumps hot brine out of the earth and runs it through a complicated series of heat exchangers to boil the butane or propane, which in turn drives the turbines.

"And there are lots of non-electrical applications," barked Norm as his plant whirred and hissed around him. "Home heating and food processing, for two examples. No one can predict the future. We haven't touched the possibilities."

T HE IMPERIAL VALLEY, a "great big jelly bowl" of miles-deep sediment topped with a thick layer of fertile soil, is one of the richest agricultural areas in the world. What made it so were the mighty Colorado River, which over the ages deposited all that rich soil as it flooded the valley and now, finally tamed, waters it; and the toil, grit, brains, and sometimes the ruthlessness of a remarkable group of men.

Cultivating the valley first became possible with completion of the Imperial Canal from the Colorado in 1901. But after 1905, when the rampaging river broke through diversion controls, the pioneer farmers struggled for decades with intermittent flood and drought. Completion of Hoover and Imperial Dams in the 1930s and the All-American Canal in 1940 finally gave the valley a dependable irrigation system, and it blossomed with fruit orchards, truck farms, and hayfields. Livestock fattened in feedlots. Pesticides, one miraculous discovery after another, slaughtered the voracious insects that had glided like pickpockets through the fields, plucking profits from farmers' wallets. The air in this low, hot place—El Centro is 52 feet below sea level, and temperatures in summer can exceed 115°F—is redolent of fertilizer, chemicals, vegetable and animal smells—the aroma of money. In a good year, it grows 800 million dollars' worth of crops on about half a million irrigated acres.

It's hard-earned money, for the farmers as well as for the sweating field workers. When Alex Abatti, the son of Italian immigrants, came home from Army service in Germany in 1957, he went to work as a field hand for $400 a month. With money saved from his wages, he bought a small piece of land of his own. Today he owns or leases thousands of acres and employs hundreds of men, paying them an average of $60 to $80 a day. "Alex is a creative capitalist, but he is also an artist as a farmer," says a friend. "He can look at the shade of blue in a field of alfalfa and know the precise moment to harvest it."

What Alex sees, more often than the blue in alfalfa, is red. He has been fighting weather, insects, government red tape, and the fickle marketplace for more than 20 years, rising at four and going home at ten at night, seven days a week, in order to make the American Dream come true for his family.

While Alex detailed his problems—how a heavy rain or a trade policy decided in Washington can threaten a year's profits; how his children, for whom he's worked the land for most of his adult life, may not want the heartache of it even if he is able to keep it—his friend Ed Chew

Contoured mazes of Arizona mines yield more copper than all the rest of the nation produces. At Kennecott's Ray operation, sectioned leach dumps form mosaics in the shadow of the Pinal Mountains. Two tons of ore-bearing rock and earth surrender a mere 17 pounds of copper; giant earth-moving machines and complex metallurgical processes keep the industry profitable.

listened with a smile of Oriental resignation. Ed was Alex's first employer; but when he came to the valley from Hong Kong at 16, an American citizen by reason of his father's earlier residence in this country, he couldn't speak a word of English. Today in El Centro, in jeans and a baseball cap, he can be mistaken for nothing but what he is: an American and a self-made man.

"Yeah," nodded Ed Chew. "In the last ten years the bureaucrats have taken all the joy out of it. But I'm not gonna let 'em broke me."

Farmers like Alex Abatti and Ed Chew, along with other southwesterners—including ORV enthusiasts and environmentalists and those who simply want to live in the sun—are caught in the slipstream of complex questions of public policy. The debates concern all of the region's limited resources: water, wilderness, minerals, timber, farmland, even homesites. But always, it seems, the most critical questions concern water.

Many of the rivers of the Southwest have been impounded, beginning with the completion of Roosevelt Dam on the Salt River in 1911, to control floods and to irrigate the Great American Desert. But by far the most controversial reclamation issue has been the distribution of the waters of the Colorado River, which passes through or borders five states and Mexico on its way to the Gulf of California.

The Colorado's water was originally apportioned by the Colorado River compact of 1922. Arizona did not ratify the compact until 1944, and at one point sent National Guard troops to the riverbanks in an attempt to prevent construction of Parker Dam. In 1952 Arizona sued California over their relative shares. Finally in 1963 the Supreme Court, after reviewing six million words of testimony in the case—one of the longest transcripts in the history of the court—allotted 2.8 million acre-feet to Arizona, 4.4 million acre-feet to California, and 300,000 acre-feet to Nevada.

The flow of the Colorado was reckoned at more than 15 million acre-feet per year by a government survey taken in 1920. But that study was made at the end of a 20-year period in which, according to tree-ring studies, there had been higher rainfall than in any similar period for at least 400 years. So the annual flow may in fact average more like 13 million acre-feet. Meanwhile Colorado has never taken its share; if ever it decides to do so, in order to develop its western slope, the water available to downriver states will diminish accordingly.

Pondering all those facts, I felt in need of a little hope; so I called on a physicist who is prepared to give the great Southwest all the pure, cheap water it will ever be able to use.

His name is John L. Hult, and he believes, with the cheery self-assurance of a scientist who has reduced a problem to a blackboard filled with equations, that desert folks will be brushing their teeth and irrigating their lettuce in 20 years with water supplied by icebergs.

In a study for the Rand Corporation and in other works, Dr. Hult has proposed floating icebergs up from Antarctica, mooring them off the California coast, melting them, and pumping the water ashore. "Iceberg water would be purer than anything we ever got out of the Colorado," Dr. Hult told me, gazing in the direction of Utopia as we chatted in his living room in Santa Monica. "Rivers as they flow gather salts. Water from the icebergs would be as pure as distilled water, for they were formed in a virtually pollution-free environment."

Dr. Hult proposes removing only bergs that *(Continued on page 194)*

Spanning Black Canyon between Arizona and Nevada, multipurpose Hoover Dam harnesses the unruly Colorado River to provide electrical power to three states and to facilitate irrigation projects downstream. When full the dam's reservoir, 110-mile-long Lake Mead, contains enough of the river's flow to submerge Pennsylvania under a foot of water. Millions of visitors, many of them braving summer infernos, cross the desert each year to vacation here.

FOLLOWING PAGES: *Water-skiers reach speeds of nearly 100 miles an hour during the international marathons on Lake Mead sponsored each September by the Las Vegas Boat & Ski Club.*

187

Lone dune buggy cruises the Imperial Sand Dunes of extreme southeastern California as afternoon fades. On fall, winter, and spring weekends, thousands of "off-road vehicle" enthusiasts spill out of the Los Angeles and San Diego areas and spread across the desert on such ORVs as trail bikes, larger motorcycles, dune buggies, and jeeps. Controversy swirls around some of their activities: Conservationists point to damage inflicted on the ecologically fragile desert, while ORV spokesmen protest the restriction of their events to limited areas of public lands. At right, the All-American Canal cuts through the dunes west of Yuma, Arizona, carrying Colorado River water from Imperial Dam to the Coachella and Imperial Valleys.

Irrigation transforms the Yuma Valley and Yuma Mesa into a year-round agricultural wonderland. With rainfall averaging less than three inches annually, farmers rely on the Colorado River and—beyond the limits of the irrigation district's canals—on wells. Rotating sprinkler systems connected to well pumps account for the circular fields. In California's Imperial Valley (right), workers pitch melons from field to truck near El Centro.

have calved off the Antarctic ice shelves, and he foresees no effect on the environment at the bottom of the globe. Australia and Chile are already investigating such sources. "We could bring up ten million acre-feet at a time—two or three years' supply for southern California."

Each berg would be hundreds of times larger than the biggest supertanker. Wouldn't that create problems? "Oh, sure," replied Dr. Hult, buoyantly. "Perhaps we could tow the bergs, wrapped in plastic to minimize melting, with submarine tugs. Perhaps we could use the temperature differential between the berg and its environment to make power. There are plenty of technological problems, but solving those would be exciting."

His smile faded. "The political problems are something else again," he said, with the mixture of sadness and frustration that afflicts that special breed, part pure thinker and part man of action. "Some people think you're just pulling their leg. But the fact is, there's enough water locked up in Antarctica to keep our desert productive—and to feed populations on waterless lands everywhere in the world."

T HE THOUGHT of Dr. Hult's icebergs, bobbing in the surf somewhere off San Diego while golden girls and boys rode their surfboards in the cool shadows, cheered me as I drove eastward. My goal was Blythe, a California town facing the Arizona border, and my idea was to take one more flight in a small plane in order to look down, in the horizontal light of the waking sun, upon the works of man.

We took off at dawn and my pilot, a full-time fireman and part-time flight instructor named Joseph F. Sheble II, assured me that I had never seen anything like what I was about to see. He was right. Joe banked the plane over the Colorado River, placid and green as a canal as it flowed among the mesas, then stood her on her right wing and pointed.

There below was an enormous human figure at least 150 feet long. Nearby was the outline of a deerlike animal. There were also winged objects and a number of concentric circles resembling archery targets. We flew on, across the river, and saw shapes we would call Maltese crosses, a tripod-like symbol, and another figure of a man.

The first group of figures, called the Blythe intaglios, was discovered in 1932 by a pilot who happened to fly over this part of the desert. The other set was detected in 1951. Even the makers of the intaglios presumably never saw them, for they are wholly visible only from the air.

Joe landed on a mesa top and we walked from the plane to one of the human figures. It was a gloomy morning, promising rain, with a pond of sunlight glimmering on a flat horizon near the Big Maria Mountains, still purple with night. Had Joe Sheble not known exactly where the intaglio was, I would have walked right over it, blind to its presence. We crouched and examined the ground. The whole figure seemed to have been made by turning over stones that were dark on one side, light on the other, to expose the pale side to the heavens. "Who did this? When? Why?" asked Joe. "There had to be a reason."

I agreed. Perhaps, I thought, it was the same reason that men in our own time have dammed rivers, made tan fields green, raised Phoenix and Tucson as heliographs to the flickering stars. Those vanished ancients may have felt the need to say, just as architect Ned Sawyer says, "Man placed this here. But his hand was skillful."

Perhaps, too, they were asking a question: Why are we here, and what is the purpose of our works?

Acknowledgments

The Special Publications Division is grateful to the individuals and organizations named or quoted in the text and to those listed here for their generous cooperation and assistance during the preparation of this book: Maxie and Patty Anderson, Dr. Donald L. Baars, John G. Babbitt, Richard L. Barnell, Dr. Lyman Benson, Leon H. Berger, Jerry Cahill, Lee Cannon, Faustin Chavez, Platt Cline, Patsy Cooley, Dr. C. Gregory Crampton, Percy Deal, Dr. David S. Dibble, Christine Dusatko, Dr. Bernard L. Fontana, Eugene and Mary Foushee, Dr. Stewart H. Fowler, Pat Foy, Marie A. Fuss, Bill Ganong, Maj. Bobby J. Good, Monica Sosaya Halford, Richard Halford, Juan Hamilton, Steve Harris, Dr. Emil W. Haury, Christopher Helms, Jeff Holladay, Carol Howard-Fitzgerald, Barbara Neel Hughes, Maj. Gen. and Mrs. Harvey J. Jablonsky, Dr. Myra Ellen Jenkins, Fred M. Johnson, Mike Kabotie, Robert A. Karges, Dr. John L. Kessell, William Guich Koock, Philip Koski, Chester Lambert, Roger Lewis, C. Laurence Linser, Amy Jo Long, Delbert D. McNealy, Ross McSwain, Saba McWilliams, David D. May, Mihran Miranian, James E. Mitchell, Kathy Morgan, Dr. John W. Morris, New Mexico State University's Department of Horticulture, Dr. James N. Norris, Professor Charles S. Peterson, George W. Ridge, Noreen M. Ross, L. W. (Budge) Ruffner, Polly Schaafsma, Albert H. Schroeder, Wade C. Sherbrooke, Dr. William V. Sliter, the Smithsonian Institution, Donald N. Soldwedel, Valerie Taliman, the Texas Wheat Producers Board, Anita Gonzalez Thomas, Dan L. Thrapp, Stanley and Virginia Throssell, Karen and Steve Trammel, Lewis D. Tyler, U. S. Army Corps of Engineers, Dr. George Watson, Dr. Marta Weigle, Col. W. C. (Doc) Wiley, Winston L. Wilson, Steven R. Wood, DeWayne Wynn.

Additional Reading

The reader may want to check the National Geographic Index for related articles and to refer to the following books: Lewis Atherton, *The Cattle Kings*; Donald L. Baars, *Red Rock Country*, The Geologic History of the Colorado Plateau; John Francis Bannon, *The Spanish Borderlands Frontier, 1513-1821*; Warren A. Beck, *New Mexico*; Charles Bowden, *Killing the Hidden Waters*; Arthur L. Campa, *Hispanic Culture in the Southwest*; Fray Angelico Chavez, *My Penitente Land*; Platt Cline, *They Came to the Mountain*, The Story of Flagstaff's Beginnings; J. Frank Dobie, *The Longhorns*; T. R. Fehrenbach, *Comanches*, The Destruction of a People; Laura Gilpin, *The Enduring Navaho*; LeRoy R. Hafen, *The Mountain Men and the Fur Trade of the Far West*, ten vols.; Cleve Hallenbeck, *Álvar Núñez Cabeza de Vaca*; Emil W. Haury, *The Hohokam*; Pauline Henson, *Founding A Wilderness Capital*: Prescott, A. T. 1864; W. Eugene Hollon, *The Great American Desert*, Then and Now; Paul Horgan, *Great River*, The Rio Grande in North American History; Norris Hundley, Jr., *Dividing the Waters*; Harry C. James, *Pages from Hopi History*; John L. Kessell, *Kiva, Cross and Crown*, The Pecos Indians and New Mexico 1540-1840; Clyde Kluckhohn and Dorothea Leighton, *The Navaho*; Howard R. Lamar, ed., *The Reader's Encyclopedia of the American West*; Paul S. Martin and Fred Plog, *The Archaeology of Arizona*; David E. Miller, *Hole-in-the-Rock*; Georgia O'Keeffe, *Georgia O'Keeffe*; Alfonso Ortiz, ed., *Handbook of the North American Indian*, vol. 9, The Southwest; Frank Gilbert Roe, *The Indian and the Horse*; Polly Schaafsma, *Indian Rock Art of the Southwest*; Paolo Soleri, *Matter Becoming Spirit*; John Upton Terrell, *Pueblos, Gods and Spaniards*; Marshall Trimble, *Arizona*; Jay J. Wagoner, *Early Arizona*; Thomas Weaver, ed., *Indians of Arizona*; Walter Prescott Webb, *The Great Plains* and *The Texas Rangers*; David J. Weber, ed., *New Spain's Far Northern Frontier*; David J. Weber, *The Taos Trappers*; Marta Weigle, *Brothers of Light, Brothers of Blood*. The reader may also wish to consult the *States and the Nation* series for individual histories of the Southwestern states, published by W. W. Norton & Company, Inc., under the auspices of the national Bicentennial of the American Revolution and the American Association for State and Local History.

Composition for *The Great Southwest* by National Geographic's Photographic Services, Carl M. Shrader, Chief, Lawrence F. Ludwig, Assistant Chief. Printed and bound by Holladay-Tyler Printing Corp., Rockville, Md. Color separations by The Beck Engraving Co., Philadelphia, Pa.; The Lanman Companies, Washington, D. C.; National Bickford Graphics, Inc., Providence, R.I.; Progressive Color Corp., Rockville, Md.; and The J. Wm. Reed Co., Alexandria, Va.

Representing life cycles of vastly different lengths, the sturdy, weathered surface of a 1,500-year-old bristlecone pine and a delicate flower of the ephemeral spiderwort form a striking contrast. Both plants appear over a wide range of the Southwest's plateau and canyon country, but the bristlecone stays at elevations above 7,500 feet.

Library of Congress CIP Data

McCarry, Charles.
    The great Southwest.

    Bibliography: p.
    Includes index.
    1. Southwest, New—Description and travel—1951-2. McCarry, Charles.
I. Mobley, George F. II. National Geographic Society, Washington, D. C. Special Publications Division. III. Title.

F787.M25   979   78-21450
ISBN 0-87044-283-X (regular binding)
ISBN 0-87044-288-0 (library binding)

Index

Illustrations references appear in **boldface**.